FLORA OF THE

T0141660

Edited by

A.R.A. GÖRTS-VAN RIJN

Series C: Bryophytes
Fascicle 1

MUSCI III

continuation of
FLORA OF SURINAME VI(I),
MUSCI II

LEUCOMIACEAE
(K. Veling)

THUIDIACEAE
(H.R. Zielman)

SEMATOPHYLLACEAE
(J. Florschütz-de Waard)

HYPNACEAE
(J. Florschütz-de Waard & K. Veling)

1996
Royal Botanic Gardens, Kew

CONTENTS

INTRODUCTION by J. Florschütz-de-Waard · · · · · · · · · · · · · · · · · · · p. 363

LEUCOMIACEAE by K. Veling · p. 365

THUIDIACEAE by H.R. Zielman · p. 371

SEMATOPHYLLACEAE by J. Florschütz-de Waard · · · · · · · · · · · · · · · p. 384

HYPNACEAE by J. Florschütz-de Waard & K. Veling · · · · · · · · · · · · p. 439

NUMERICAL LIST OF ACCEPTED TAXA · p. 463

COLLECTIONS STUDIED · p. 465

INDEX TO SYNONYMS · p. 476

KEY TO THE GENERA OF MOSSES OF THE GUIANAS
by J. Florschütz-de Waard · p. 480

INTRODUCTION

by

J. FLORSCHÜTZ-DE WAARD[1]

With this third part of the Moss Flora of Suriname the treatment of the families is completed. Parts 1 and 2 have been published in Vol. VI, 1 of the Flora of Suriname (Florschütz 1964, Florschütz-de Waard 1986). In all three parts the collections from Guyana and French Guiana, so far as available, have also been included since the flora of the three countries is essentially similar. For this reason it has been decided to publish part 3 in the "Flora of the Guianas". The pages and illustrations are numbered in the sequence started in the Flora of Suriname. The terminology used in the descriptions follows Ireland (1982)[2].

A key to the genera occurring in the Guianas is added at the end of this part. Since the appearance of Parts 1 and 2 new species, genera and families have been collected in the region, especially at higher altitudes in Guyana, which were neglected previously. Most of these additional species have been included in the recent checklist of the Guianas (Florschütz-de Waard 1990, Boggan et al. 1992)[3]. These additions will be treated in a fourth, supplementary part, which will provide an updated overview of the moss flora of the Guianas.

I am grateful to H.R. Zielman who assisted in treating the family Thuidiaceae and spent much of his spare time making a thorough study of this group. I also wish to thank my student K. Veling for his contributions to the families Leucomiaceae and Hypnaceae. I am indebted to the Curators of the herbaria cited for the loan of specimens, and to Dr. A. Newton for critically reading the manuscript and for valuable comments on the general key to the genera.

[1] Herbarium Division, Department of Plant Ecology and Evolutionary Biology. Heidelberglaan 2, 3584 CS Utrecht, The Netherlands.

[2] Ireland, R.R., 1982. Moss Flora of the Maritime Provinces. Nat. Mus. Nat. Sc. Ottawa, Bot. 13, glossary: 702-727.

[3] Florschütz-de Waard, J., 1990. A Catalogue of the Bryophytes of the Guianas II. Musci. Tropical Bryology 3: 89-104.

Boggan J. et al., 1992. Checklist of the plants of the Guianas. Washington, DC: Biological Diversity of the Guianas Program.

LEUCOMIACEAE

by

K. VELING[4]

Plants with creeping stems, without central strand, irregularly branched, branches prostrate. Leaves ecostate, ovate to lanceolate, acute or acuminate; leaf cells lax, smooth, alar cells not differentiated.

Capsules with rostrate operculum, exothecial cells collenchymatous; peristome double, exostome teeth furrowed and striate in basal part, endostome with broad, keeled segments on a high basal membrane. Calyptra cucullate.

Note: Brotherus (1909) established the family for *Leucomium* and *Vesiculariopsis*. Originally Mitten (1869) placed *Leucomium* in the Hypnales in consequence of its similarity to *Vesicularia* (Hypnaceae). Important differences with *Vesicularia* are: the lack of stem differentiation, the absence of pseudoparaphyllia, the long-rostrate operculum and the furrowed exostome teeth. Fleischer (1923) was the first to place the Leucomiaceae (including *Leucomium* and *Rhynchostegiopsis*) in the Hookeriales. Most floristic treatments consider *Leucomium* a hookerioid genus, but the position of the family either in the Hookeriales or in the Hypnales has remained subject to discussion. Allen (1987) demonstrated a nearly complete character agreement of *Leucomium* with *Hookeria*, except for the cucullate calyptra. He concluded a near relationship of the family to the Hookeriaceae, only separated by the calyptra characters and he supported Robinson's view (1986) that the Leucomiaceae take an intermediate position between Hookeriaceae and Sematophyllaceae.

Distribution: only one genus in the Guianas.

LITERATURE

Allen, B.H. 1987. A revision of the genus Leucomium (Leucomiaceae). Mem. New York Bot. Gard. 45: 661-677.

Brotherus, V.F. 1909. In: A. Engler & K. Prantl. Die natürlichen Pflanzenfamilien 1(3), Leipzig. Leucomiaceae: 1095-1098.

Fleischer, M. 1923. Die Musci der Flora von Buitenzorg 4: 1105-1109.

Mitten, W. 1869. Musci Austro-Americani. J. Linn. Soc. Bot. 12: 500-503.

Robinson, H. 1986. On the relationships of the Hookeriaceae. Bryol. Times 35: 2-3.

[4] Herbarium Division, Department of Plant Ecology and Evolutionary Biology. Heidelberglaan 2, 3584 CS Utrecht, The Netherlands.

1. **LEUCOMIUM** Mitt., J. Linn. Soc. Bot. 10: 181. 1868.
   Type: L. debile (Sull.) Mitt. (Hookeria debilis Sull.)

Slender to medium-sized plants growing in loose mats. Stems and branches elongate, densely foliate; pseudoparaphyllia absent. Stem- and branch leaves not differentiated, ovate to lanceolate, variably dimorphic; costae absent; apex acute or acuminate; margin entire. Leaf cells lax, rhomboidal to elongate-hexagonal, thin- or firm-walled.
Synoicous or autoicous. Seta elongate, smooth or slightly rough below the capsule; capsule inclined to pendulous, operculum slenderly rostrate; exostome teeth furrowed, densely striate below, papillose at the tip, at ventral side with projecting lamellae, endostome with high basal membrane and broad, keeled segments, smooth or minutely papillose, cilia absent. Calyptra narrowly cylindric.

Distribution: pantropical.

KEY TO THE SPECIES

1 Seta 8-20 mm long; capsule to 1mm long (without operculum). Leaves 1.3-2 mm long, acute or acuminate $\cdots\cdots\cdots\cdots\cdots\cdots\cdots\cdots$ *2. L. strumosum*
  Seta 25-30 mm long; capsule 1-2 mm long (without operculum). Leaves 1.5-3 mm long, with a long piliferous (often hyaline) acumen $\cdots$ *1. L. steerei*

1. **Leucomium steerei** B.H. Allen & Veling, Mem. New York Bot. Gard. 45: 674. 1987. Type: Venezuela, Bolívar, Chimanta Massif, Steyermark 75964A (FH).   – Fig. 142.

Medium-sized, green or orange-tinged plants. Stems creeping, branches short, prostrate or ascending. Leaves ovate-lanceolate to lanceolate, long-acuminate with often hyaline hairpoint, 1.5-3.1 mm long, 0.3-0.8 mm wide; leaf cells firm-walled, elongate-rhomboidal or elongate-hexagonal, sometimes nearly linear, 90-180 μm long and 10-20 μm wide, basal cells shorter.
Autoicous? Perichaetial leaves ovate-lanceolate, long-acuminate, to 3 mm long; perigonia not seen. Seta firm, to 30 mm long; capsule ovoid-cylindric, 1-2 mm long, inclined to horizontal; peristome with the characters of the genus, exostome teeth firm, 700 μm long, endostome segments papillose at apex.

Distribution: Guayana highlands. Endemic.

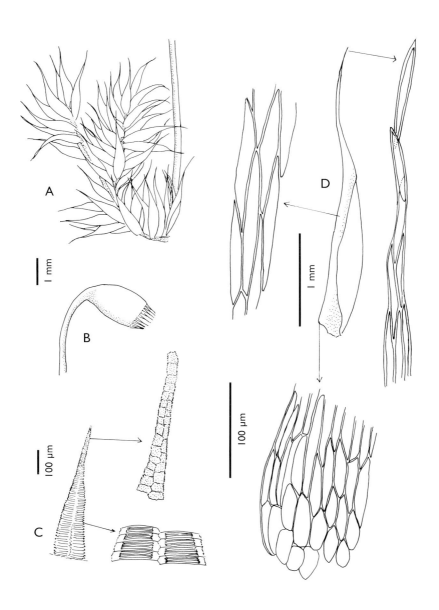

Fig. 142. *Leucomium steerei*: A. part of stem with branch and perichaetium; B. capsule; C. exostome tooth; D. leaf (Gradstein 5421).

Ecology: Epiphytic in humid, "mossy" forest at higher altitudes.

Specimens examined: Guyana: Upper Mazaruni District, N. slope of Mt. Roraima, Alt. 1200-1600 m, Gradstein 5286, 5421 (U).

Note: The extraordinary size of the sporophyte and the occurrence at higher altitudes are the determinant characters of *L. steerei*. The gametophytes are sometimes hard to distinguish from specimens of *L. strumosum* (expression *strumosum*) which also have lanceolate leaves with slenderly acuminate and occasionally even hairpointed apex. But in that case the leaf cells are usually thin-walled and thus the combination of hairpointed leaves and firm-walled cells are decisive for *L. steerei*.

2. **Leucomium strumosum** (Hornsch.) Mitt., J. Linn. Soc. Bot. 12: 502. 1869. – *Hookeria strumosum* Hornsch., Fl. Bras. 1(2): 69. 1840. – *Hypnum strumosum* (Hornsch.) C. Müll., Syn. 2: 238. 1851. Lectotype (Allen 1987): Brazil, prope Tijucam, Olfers s.n. (BM).
– Fig. 143.

Leucomium cuspidatum (C. Müll.) Jaeg., Ber. S. Gall. Naturw. Ges. 1877-78: 275. 1880. – *Acosta cuspidata* C. Müll., Linnaea 21: 192. 1848. – *Hypnum cuspidatifolium* C. Müll., Syn. 2: 237. 1851. Type: Suriname, Paramaribo, Febr. 1844, Kegel 990 (GOET).
Leucomium guianense C. Müll., Malpighia 10: 515. 1896. Type: Guyana, Marshall Falls, Mazaruni River, Quelch s.n., hb. Levier 1267 (BM).

For complete synonymy see Allen (1987).

Small to medium sized, pale green plants. Stems creeping, 2-3 cm long, irregularly branched, branches elongate, prostrate or slightly ascending. Leaves erect-spreading or more or less complanate, often homomallous, ovate-lanceolate, gradually to abruptly acuminate, 1.3-2 mm long, 0.4-0.6 mm wide, margin entire; midleaf cells large, thin-walled or firm-walled, elongate-rhomboidal or elongate-hexagonal, sometimes nearly linear, 100-200 µm long and 10-40 µm wide, basal cells shorter and wider, alar cells not differentiated.
Synoicous. Perichaetial leaves lanceolate, to 1.5 mm long, gradually long-acuminate. Seta 8-14 mm long; capsule inclined to pendent, ovoid-cylindric, often slightly strumose at base, 0.5-0.7 mm long and 0.3-0.5 mm wide, operculum conic, long-rostrate; peristome with the characters of the genus, exostome teeth usually strongly inflexed when wet, to 300 µm long, endostome segments of equal length, erect when wet. Calyptra narrowly cylindric, often with scattered hairs on the base.

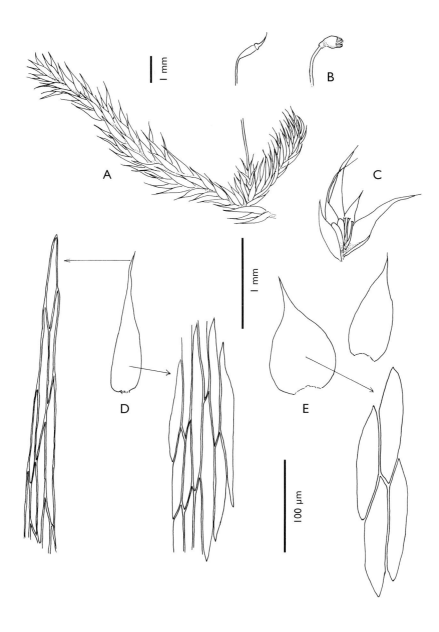

Fig. 143. *Leucomium strumosum*: A. part of stem with branch and immature capsule; B. mature capsule; C. perichaetium; D. leaf (expr. *strumosum*); E. leaf (expr. *mosenii*). (A-D: Florschütz 1432; E: Florschütz-de Waard & Zielman 5470)

D i s t r i b u t i o n : pantropical.

E c o l o g y : A common species in the understory of lowland rainforest, epiphytic, mostly on tree bases, but also on decaying wood; occasionally terrestrial or epilithic near streams.

S e l e c t e d   s p e c i m e n s : Guyana: Upper Mazaruni district, Jawalla, Alt. 500 m, Gradstein 4865 (U); Mabura Hill, Alt. 0-50 m, Cornelissen & ter Steege 106 (U). Suriname: Saramaca River, trail from Paka-paka to Ebbatop, Florschütz 1432 (expr. *strumosum*) (U); Area of Kabalebo Dam project, NW of road km 39.5, Florschütz & Zielman 5237 (U). French Guiana: Haut Maroni, confluent de l'Itany et du Marouini, Cremers 5153 (CAY, U); Crique Cabaret, Bassin de l'Oyapock, Cremers 9991 (expr. *mosenii*) (CAY, U).

N o t e : *L. strumosum* is easily recognized by the lax areolation and the lack of costae. It resembles *Vesicularia vesicularis*, but the latter is usually much more noticeably pinnate. The cell size and the capsule with slenderly rostrate operculum and furrowed peristome teeth are good characters to distinguish the two species.
L. strumosum is variable in the following characters: the leaf dimorphism, the leaf shape and the cell shape. Allen (1987) distinguished 3 expressions of which one (expr. *compressum*) is characterized by dimorphic leaves. In the Guianan collections, however, this character is rarely seen. Expr. *mosenii* has ovate or oblong-ovate and short-acuminate leaves with usually firm-walled, hexagonal leaf cells. Expr. *strumosum* is characterized by lanceolate or ovate-lanceolate leaves; the leaf cells are more elongate and narrower, sometimes nearly linear. A gradual variation between these extreme expressions makes it impossible to distinguish varieties.

THUIDIACEAE

by

H.R. Zielman[5]

Small to medium-sized plants. Stems creeping, uni- to tripinnate; paraphyllia simple or branched, papillose. Leaves distinctly dimorphous, stem leaves cordate-triangular, leaves of ultimate branches ovate; costa single, well-developed. Leaf cells hexagonal to short-rectangular, papillose.

Autoicous or dioicous. Inner perichaetial leaves larger than stem leaves, with slender acumen, usually with a more lax aereolation. Seta elongate, smooth or papillose; capsule horizontal or inclined; operculum mostly rostrate; peristome double, exostome teeth striolate below, papillose above, endostome of about equal length as exostome, segments from a high basal membrane, keeled, cilia usually present. Calyptra cucullate, smooth.

N o t e s : An important character in this family is the extent of branching. Stems (with the gametangia and rhizoids) with a single row of branches on both sides of the axis are unipinnate. If these branches have a second row of branches the plants are bipinnate, etc. Leaf lengths are measured on leaves at the middle of branches. Another character that can be useful is the aspect of the costa at the dorsal side of the leaf. The costa can be very prominent throughout or just fade away in the leaf surface at distal end.

Buck & Crum (1990) divided this family in two subfamilies: the *Thuidioideae*, with one genus (*Thuidium*) including the large to medium-sized, dioicous species, and the new subfamily *Cyrtohypnoideae*, combining the small, autoicous species in *Cyrtohypnum*.

LITERATURE

Buck, W.R. & H. Crum, 1990. An evaluation of familial limits among the genera traditionally alligned with the Thuidiaceae and Leskeaceae. Contr. Univ. Michigan Herb. 17: 55-69.

Crum, H. & L.E. Anderson, 1981. Mosses of Eastern North America (2). Columbia Univ. Press. New York.

Crum, H. & W.R. Buck. 1994. In: A.J. Sharp, H. Crum & P.M. Eckel, The moss flora of Mexico. Mem. New York Bot. Gard. 69. Thuidiaceae: 873-886.

[5] Herbarium Division, Department of Plant Ecology and Evolutionary Biology. Heidelberglaan 2, 3584 CS Utrecht, The Netherlands.

Crum, H. & W.C. Steere, 1957. The mosses of Porto Rico and the Virgin Islands. Sci. Surv. Porto Rico and the Virgin Islands 7(4). Thuidiaceae: 556-559.

Crum, H. & W.C. Steere, 1958. A contribution to the bryology of Haiti. Am. Midl. Nat. 60 (1): 1-51.

Mitten, W., 1869. Musci Austro-Americani: J. Linn. Soc. Bot. 12. Thuidium: 572-580.

Müller, C., 1851. Synopsis Muscorum Frondosorum II. Berlin.

Müller, C., 1896. Musci nonnulli novi Guianae Anglicae. Malpighia 10: 512-520.

KEY TO THE GENERA

1 Plants slender, first order branches 5 mm or less long. Paraphyllia usually simple, occasionally weakly branched. Leaf cells with small papillae on both surfaces. Autoicous. Seta often rough · · · · · · · · · · · · *1. Cyrtohypnum*
   Plants medium-sized, first order branches at least 5 mm long. Paraphyllia strongly branched. Leaf cells with rather large papillae at abaxial surface. Dioicous. Seta smooth · *2. Thuidium*

1. **CYRTOHYPNUM** (Hampe) Hampe & Lor., Bot. Zeit. 27: 455. 1869. – *Hypnum* subgen. *Cyrto-hypnum* Hampe, Flora 50: 78. 1867. – *Thuidium* subgen. *Microthuidium* Limpr., Krypt.-Fl. Deutschl. ed. 2,4: 822. 1895.
   Type: C. brachythecium Hampe & Lor. (Hypnum brachythecium Hampe & Lor.)

Slender plants, branching unipinnate to tripinnate; paraphyllia often scarce, usually unbranched, smooth below, papillose above. Stem leaves cordate-triangular, acuminate, often plicate, margins recurved, branch leaves ovate or oblong, apex acute to rounded, margins plane; cells of stem and branch leaves pluripapillose on both surfaces, papillae minute. Autoicous. Perichaetial leaves long-acuminate, with or without cilia, cells elongate, smooth; seta elongate, often rough; capsule horizontal; peristome with the characters of the family.

D i s t r i b u t i o n : mainly tropical.

KEY TO THE SPECIES

1 Branching unipinnate. Stem leaves not larger than branch leaves ·········
·· *1. C. involvens*
  Branching bi- or tripinnate. Stem leaves larger than branch leaves ······· 2

2 Plants lax, branches catenulate with strongly curved, semi-circinate leaves
  when dry; perichaetial leaves not or scarcely ciliate·················
·· *2. C. scabrosulum*
  Plants rigid, branch leaves loosely imbricate, incurved when dry; perichaetial
  leaves strongly ciliate ··························· *3. C. schistocalyx*

1. **Cyrtohypnum involvens** (Hedw.) W.R. Buck & Crum., Contr. Univ. Michigan Herb. 17: 66. 1990. – *Leskea involvens* Hedw., Spec. Musc. 218. 1801. – *Thuidium involvens* (Hedw.) Mitt., J. Linn. Soc. Bot. 12: 575. 1869. Type: Jamaica, Swartz s.n. (BM).   – Fig. 144.

Slender, creeping, yellowish green plants, forming flat mats. Stems unipinnately branched, to 4 cm long; paraphyllia very scarce, 2-4 cells long, seldom papillose; pseudoparaphyllia narrowly triangular or filiform, to 120 μm long; stem leaves little differentiated, having the same size as branch leaves, ovate-triangular, apex acuminate. Branches 2-4 mm long; branch leaves distant, ovate, curved when dry, complanate-spreading when moist, slightly asymmetric, 0.4-0.55 mm long and 0.2-0.3 mm wide, apex acute, margin crenulate, plane or narrowly recurved at one side, costa extending 3/4-7/8 of leaf length, dorsal cells linear, smooth, occasionally prorate; leaf cells rounded-quadrate to short-rectangular, 8-15 μm long and 7-12 μm wide, papillae very indistinct.
Autoicous. Perigonial leaves broadly ovate, concave, ca. 0.4 mm long. Perichaetial leaves triangular, with long acumen, to 1.7 mm long and 0.4 mm wide, margin entire or serrulate, without cilia. Seta 8-11 mm long, rough with ca. 20 μm high mamillae; capsule short-cylindric, 0.7 mm long, operculum long-rostrate; peristome teeth 250 μm long, endostome segments papillose, cilia single.

Distribution: Southern U.S.A., West Indies, Central and tropical South America, Central Africa (ssp. *thomeanum*).

Ecology: On decaying wood or terrestrial. Apparently rare in the Guianas; not known from Guyana or French Guiana.

Specimens examined: Suriname: Kabalebo River, N side of Avanavero Falls, Florschütz 2201 (U); Gros savanna, in low forest, Florschütz-de Waard 4917 (U).

N o t e : This species can easily be recognized by the unipinnate stems with hardly any paraphyllia and by the distantly foliose branches. The branch leaves have a costa in which at dorsal side elongated to linear cells are clearly visible throughout its length, in contrast to the two

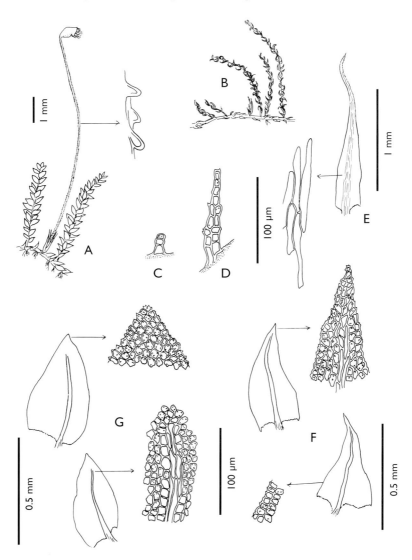

Fig. 144. *Cyrtohypnum involvens*: A. stem portion with branches and capsule (moist); B. habit (dry); C. paraphyllium; D. pseudoparaphyllium; E. perichaetial leaf; F. stem leaves; G. ultimate branch leaves. (A-B: Florschütz 2201; C-G: Florschütz-de Waard 4917).

other species in which the costa at distal end seems to be covered by quadrate to short-rectangular laminal cells. This may be helpful to distinguish poorly developed plants of *C. scabrosulum* from *C. involvens*. In addition the stem leaves in *C. involvens* are hardly larger than the branch leaves.

2. **Cyrtohypnum scabrosulum** (Mitt.) W.R. Buck & Crum, Contr. Univ. Michigan Herb. 17: 67. 1990. – *Thuidium scabrosulum* Mitt., J. Linn. Soc. Bot. 12: 574. 1869. Type: sine loc., Humboldt s.n., hb. Hooker 40 (BM).   – Fig. 145.

Thuidium verrucipes C. Müll., Malpighia 10: 519. 1896. syn. nov. Type: Guyana, Marshall Falls, Quelch s.n., hb. Levier 1293 (BM).

Slender, green to yellowish green plants, growing in loose mats; branching bipinnate, occasionally irregularly tripinnate, branchlets catenulate. Stems 2-7 cm long, often arcuate; paraphyllia numerous, with 3-5(-8) pluripapillose cells; pseudoparaphyllia filiform or narrow-triangular, 80-200 µm long. Stem leaves broad-triangular, flexuose when dry, spreading when moist, 0.2-0.8 mm long and 0.2-0.45 mm wide, apex acuminate, margin crenulate, narrowly recurved, costa extending to leaf apex; leaf cells rounded-quadrate to hexagonal, 7-18 µm long and 6-9 µm wide, papillae minute. First order branches usually 2-4 mm long, leaves intermediate in size and shape between stem leaves and leaves of ultimate branches. Leaves of ultimate branches ovate to oblong, curved to semi-circinate when dry, more or less complanate when moist, 0.15-0.3 mm long and 0.09-0.16 mm wide, apex rounded to acute, margin crenulate, plane, costa extending 2/3-3/4 of leaf length, dorsally protruding, dorsal epidermis cells generally quadrate and pluripapillose in distal part; leaf cells rounded-quadrate, 7-12 µm long and 6-11(-13) µm wide.
Autoicous. Perigonial leaves ovate, 0.3-0.5 mm long, abruptly acuminate, concave. Perichaetial leaves ovate-lanceolate with long-excurrent costa, 0.8-2.5 mm long; margin serrulate, without cilia or occasionally with a few short, straight cilia. Seta 10-16 mm long, rough with ca. 20 µm high mamillae; capsule inclined to horizontal, short-cylindric, slightly curved, ca. 1 mm long, operculum conic, rostrate; peristome teeth 300-360 µm long, endostome segments perforated, cilia single.

D i s t r i b u t i o n : West Indies, Central and tropical South America.

E c o l o g y : Epiphytic in the forest canopy or on decaying wood in open areas of the forest, occasionally terrestrial. Rather common in Suriname and French Guiana, apparently rare in Guyana.

376

Selected specimens: Guyana: Rupununi Distr., foothills of NW Kanuku Mts. near Moco-Moco village, Alt. 100 m, Maas et al. 3850 (U). Suriname: Saramaca River, trail to Ebba top (van Asch van Wijck Mts.).

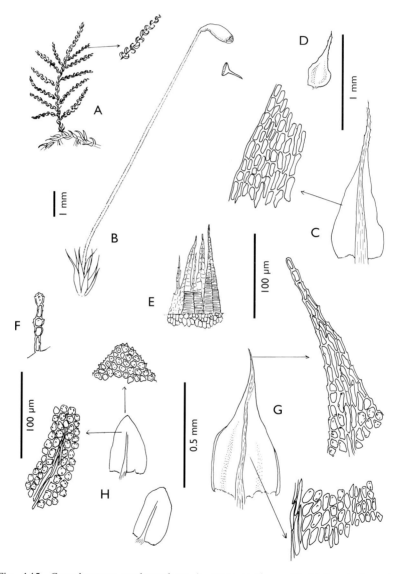

Fig. 145. *Cyrtohypnum scabrosulum*: A. stem portion with bipinnate branch (dry); B. perichaetium with capsule; C. perichaetial leaf; D. perigonial leaf; E. part of peristome; F. paraphyllium; G. stem leaf; H. ultimate branch leaves. (Florschütz-de Waard & Zielman 5564).

Florschütz 1259 (U); W bank Kabalebo River, near ferry in road to Amatopo, Florschütz-de Waard & Zielman 5353 (U). French Guiana: Sommet Tabulaire face Nord, Alt. 600-700 m, Cremers 6839 (CAY, U); Saül, primary forest near Boeuf Mort, Alt. 200-300 m, Aptroot 15216 (U).

N o t e s : Examination of the type collection of *Thuidium verrucipes* described by Müller (1896) from Guyana, proved that this species is identical with *Cyrtohypnum scabrosulum*. It is the most common *Cyrtohypnum* species in the Guianas. It is easily recognized by its bipinnate habit and the catenulate branches. The curved, distant branch leaves give the dry plants a superficial resemblance to *C. involvens*. However, *C. scabrosulum* is predominantly bipinnate. See also the note under *C. involvens*.

3. **Cyrtohypnum schistocalyx** (C. Müll.) W.R. Buck & Crum, Contr. Univ. Michigan Herb. 17: 67. 1990. – *Hypnum schistocalyx* C. Müll., Syn. 2: 691. 1851. – *Thuidium schistocalyx* (C. Müll.) Mitt., J. Linn. Soc. Bot. 12: 575. 1869. Type: Nicaragua, Matagalpa in Segovia, Oersted s.n., Dec. 1847 (BM). – Fig. 146.

Slender plants growing in dark green mats. Stems to 10 cm long, bipinnately branched; paraphyllia very abundant, 2-4(-8) cells long; pseudoparaphyllia triangular or filiform, 50-120 µm long. Stem leaves erect, cordate-triangular, 0.25-0.5 mm long and 0.2-0.3 mm wide, apex acute or acuminate, margin crenulate or serrulate, plane or narrowly recurved, costa extending 3/4-4/5 of leaf length, rough at back towards apex; leaf cells rounded-quadrate to short-rectangular, 8-13 µm long and 5-10 µm wide. First order branches 3-5 mm long, leaves intermediate between stem leaves and leaves of branchlets. Ultimate branch leaves ovate to oblong, loosely imbricate, incurved when dry, subcomplanate when moist, sometimes slightly asymmetric, 0.14-0.20 mm long and 0.07-0.12 mm wide, apex rounded-acute, margin crenulate, plane, costa extending 2/3-4/5 of leaf length, crested dorsally, at apical end covered with laminal cells; leaf cells rounded-quadrate, 6-10 µm long.
Autoicous. Perigonial leaves ovate-triangular, 0.25-0.5 mm long, apex acute, margin entire or dentate near apex. Perichaetial leaves ovate-lanceolate, to 1.5 mm long, acuminate, often flexuose, serrate in upper part, with long, flexuose cilia, costa excurrent. Seta 12-14 mm long, rough with 15-20 µm high mamillae; capsule inclined, 0.8 mm long, short-cylindric, operculum rostrate; peristome teeth 320-400 µm, endostome segments 320 µm, cilia single, split at apex.

D i s t r i b u t i o n : Southern U.S.A., West Indies, Central and tropical South America.

Ecology: Epiphytic or on decaying wood, occasionally on rocks. Not common; not known from Guyana.

Selected specimens: Suriname: Paramaribo, Cultuurtuin, on bark of *Couratari* tree, Van Looy 43 (U); Area of Kabalebo Dam project, road

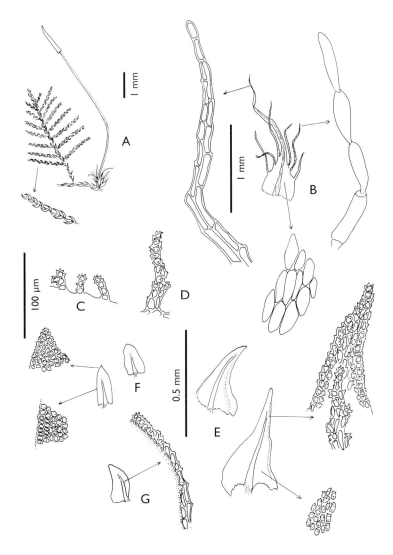

Fig. 146. *Cyrtohypnum schistocalyx*: A. stem portion with bipinnate branch (dry) and perichaetium with capsule; B. perichaetial leaf; C. paraphyllia; D. pseudoparaphyllium; E. stem leaves; F. ultimate branch leaves; G. ultimate branch leaf with costa in side view. (Florschütz 4538).

to Amatopo km 212.5, Florschütz-de Waard & Zielman 5777 (U). French Guiana: Saül, Circuit Petit Boeuf Nort, Cremers 4181 (CAY, U); Eaux Claires, 5 km N of Saül, Florschütz-de Waard 5854 (U).

N o t e : The white perichaetia of this species are very conspicuous when young, but in older perichaetia the cilia are often eroded. The plants have a more rigid appearance than *C. scabrosulum* because of the patent-spreading, less flexuose branchlets. An additional character with some overlap is the number of branchlets per branch: *C. scabrosulum* has 3-14 branchlets and *C. schistocalyx* 9-18.

2. THUIDIUM Schimp., Bryol. Eur. 5: 157. 1852.

Type: T. tamariscinum (Hedw.) Schimp. (Hypnum tamariscinum Hedw.)

Medium-sized to rather robust plants, forming dense mats. Stems uni-pinnate- to tripinnate-branched, often arched; paraphyllia abundant, strongly branched. Stem leaves cordate-triangular, acuminate, plicate, margins recurved; branch leaves ovate, apex acute, margins plane. Leaf cells with tall papillae at abaxial leaf surface.
Dioicous. Perichaetial leaves long-acuminate, costa percurrent, cells elongate, smooth, marginal cilia often present. Seta long, smooth; capsule inclined or horizontal; peristome with the characters of the family.

D i s t r i b u t i o n : world-wide.

KEY TO THE SPECIES

1 Leaf cells unipapillose · *1. T. peruvianum*
  Leaf cells pluripapillose · *2. T. tomentosum*

1. **Thuidium peruvianum** Mitt., J. Linn. Soc. Bot. 12: 578. – *Thuidium delicatulum* (Hedw.) B.S.G. var. *peruvianum* (Mitt.) Crum, Bryologist 87: 211. 1984. Type: Peru, Mathews s.n. (NY).   – Fig. 147.

Plants growing in dense mats. Stems 5-12 cm long, branching bipinnate or irregularly tripinnate; paraphyllia spinose-papillose; pseudoparaphyllia foliose, to 0.5 mm long, often laciniate. Stem leaves cordate-triangular, 2.5-2.9 mm long, 0.8-1.0 mm wide, strongly plicate, erecto-patent when dry, patent when moist, obliquely inserted, apex long-acuminate with

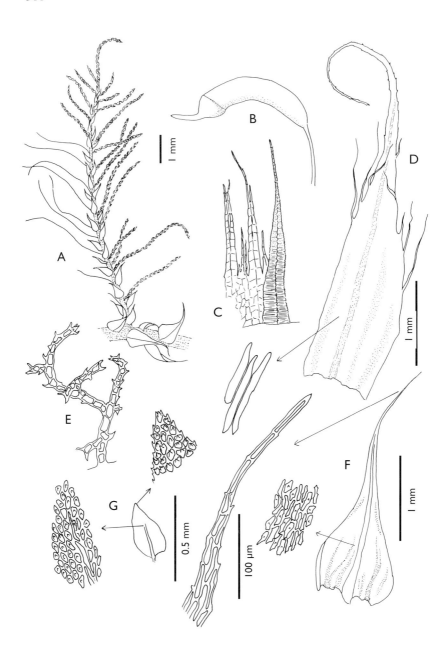

Fig. 147. *Thuidium peruvianum*: A. stem portion with bipinnate branch (dry); B. capsule; C. exostome tooth and part of endostome; D. inner perichaetial leaf; E. paraphyllium; F. stem leaf; G. ultimate branch leaf. (Appun s.n.).

filiform acumen 3-6 cells long, margin denticulate, more strongly so towards apex, recurved in lower part, costa reaching acumen, ending well below apex; leaf cells irregular-rectangular, 17-32 µm long and 6-9 µm wide, incrassate, with one low central papilla, in apex linear, smooth. First order branches 1-1.5 cm long, leaves lanceolate, with acute apex, margin plane, denticulate, costa extending 2/3-4/5 of leaf length. Leaves of ultimate branches ovate, 0.2-0.3 mm long and 0.13-0.20 mm wide, apex acute, margin dentate, plane, costa to midleaf, dorsal cells elongated; leaf cells rounded-rectangular, 10-14 µm long and 7-10 µm wide, incrassate, with a very prominent curved papilla, apical cell crowned with papillae.

Dioicous. Perigonia not seen. Perichaetial leaves plicate, narrowly triangular, long-acuminate, to 5.5 mm long, margin ciliate, strongly dentate in upper part. Seta to 5 cm long; capsule arcuate, to 4 mm long, exclusive of the rostrate operculum; peristome teeth ca. 900 µm long, endostome segments papillose, often split at the top, cilia 1-3.

Distribution: Central and tropical South America.

Ecology: On the label of the only collection from the Guianas no habitat data are mentioned. Outside the Guianas *T. peruvianum* is reported from soil and rock at higher altitudes.

Specimen examined: Guyana: Appun s.n. (BM).

Notes: This species is separated from the wide-spread *Thuidium urceolatum* Lor. (*T. acuminatum* Mitt.), that also has unipapillose leaf cells, by the conspicuous long, filiform acumen of the stem leaves. *T. peruvianum* belongs to the affiliation of *T. philibertii* Limpr., both being considered varieties of *T. delicatulum* (Hedw.) B.S.G. by some authors, e.g. Crum & Buck (1994) and Crum & Anderson (1981). *T. peruvianum* differs from *T. philibertii* in the ciliate perichaetial leaves, from *T. delicatulum* in the long-acuminate stem leaves, and from both in its larger dimensions. Awaiting a critical revision of the *Thuidium philibertii* complex that also considers other tropical taxa I prefer a conservative position.

2. **Thuidium tomentosum** Schimp., Mem. Soc. Sci. Nat. Cherbourg 16: 237. 1872. Type: Mexico, Orizaba, Fr. Müller Bescherelle s.n. (NY).
– Fig. 148.

Thuidium antillarum Besch., Ann. Sci. Nat. Bot. ser. 6,3: 244. 1876. Type: Guadeloupe, Beaupertuis s.n. (BM).

382

Green to yellowish green, rather robust plants growing in dense mats with conspicuously curved branches when dry; branching bipinnate to irregularly tripinnate. Stems often arcuate, 5-7 cm long; paraphyllia cells quadrate, with several sharp papillae except on the basal ones; pseudoparaphyllia filiform to narrow-triangular, to 0.5 mm long; leaves

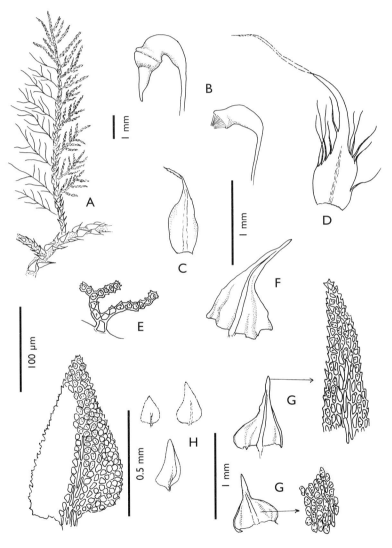

Fig. 148. *Thuidium tomentosum*: A. stem portion with bi- to tripinnate branch (dry); B. capsules; C. inner perigonial leaf; D. inner perichaetial leaf; E. paraphyllium; F, G. stem leaves; H, I. ultimate branch leaves. (A-E, G-I: Lanjouw & Lindeman 2669; F: Florschütz-de Waard & Zielman 5050).

ovate-triangular, 0.6-1.2 mm long, 0.3-0.7 mm wide, strongly plicate, erect to appressed and slightly homomallous when dry, erecto-patent when moist, apex acuminate, margin crenulate or serrulate, more strongly so near apex, recurved in lower part, costa ending below apex, often with a few short paraphyllia at the base; leaf cells rounded-quadrate to rectangular, 5-10 µm long and 8-16(-20) µm wide, with 2-4 papillae at abaxial surface, at extreme leaf base without papillae, apical cell crowned with papillae. First order branches 5-15 mm long, branchlets more or less catenulate; leaves of ultimate branches ovate, concave, 0.12-0.3 mm long and 0.07-0.18 mm wide, apex acute, margin serrulate, costa protruding at dorsal leaf side, extending 1/2-3/4 of leaf length, occasionally almost lacking, dorsal cells elongate, smooth or prorate; leaf cells rounded-hexagonal, 6-14 µm long and 6-10 µm wide, incrassate, with (1-)2-4 papillae, concentrated over centre of lumen, often curved in distal direction, apical cell truncate, crowned with papillae.

Dioicous. Outer perigonial leaves small, broad-ovate, with reflexed apiculus, inner perigonial leaves to 1 mm long, abruptly acuminate. Perichaetial leaves ovate-triangular, to 3.5 mm long including the flexuose apex (to 2 mm long), margin ciliate or laciniate in upper part, cilia long and flexuose. Seta to 4.5 cm long; capsule inclined, arcuate, 1-2 mm long, exclusive of the conic-rostrate operculum; peristome teeth 500-650 µm long, endostome segments not perforated, cilia single.

D i s t r i b u t i o n : West Indies, Central and tropical South America.

E c o l o g y : On stones and on bark of trees, occasionally on decaying wood or terrestrial, generally at higher altitudes. Common throughout the area.

S e l e c t e d s p e c i m e n s : Guyana: Kaieteur Falls, Tutin 538 (BM); Mt. Latipu, ca. 8 km N of Kamarang, Alt. 600 m, Gradstein 5564 (U). Suriname: Nassau Mts., plateau, Alt. ca. 500 m, Lanjouw & Lindeman 2669 (U); Tafelberg, savanna no. VIII, Maguire 24433 (NY, U). French Guiana: Mt. de l' Inini, Feuillet 3896 (CAY, U); Station de Nouragues, Bassin de l' Approuage, Arataye, Cremers 11037 (CAY, U).

N o t e : *Thuidium antillarum* and *T. acuminatum* frequently have been reported for the Guianas. *T. acuminatum* Mitt. (1869), placed in synonymy of *T. urceolatum* Lor. by Crum & Steere (1957), is a species with uni-papillose leaf cells, which has not yet been collected in the Guianas. *T. antillarum* has also been synonymized with *T. urceolatum* by Crum & Steere (1958). However, the leaf cells in the type specimen of *T. antillarum* are pluripapillose and it proves to be identical with *T. tomentosum*; recently it has been placed in synonymy of this species (Crum & Buck 1994).

SEMATOPHYLLACEAE

by

J. Florschütz-de Waard[6]

Slender to medium sized plants growing in flat or rough mats. Stems creeping, irregularly or pinnately to bipinnately branched, branches horizontal or ascending; pseudoparaphyllia foliose, filamentous or lacking. Leaves oval, ovate, oblong or lanceolate, costa short and double, usually indistinct; leaf cells smooth or papillose, rhomboidal to linear, alar cells differentiated, usually inflated in the basal row.
Capsule erect, inclined or pendulous, operculum often rostrate, exothecial cells collenchymatous or incrassate along longitudinal walls; peristome single or double, exostome teeth slender and pale or strong, brown and thickened by high transverse lamellae on inner side, endostome absent or reduced to a low basal membrane with slender segments or well-developed with a high basal membrane with broad, keeled segments and cilia. Calyptra cucullate.

N o t e : From the description it is clear that this family is rather heterogeneous in gametophytic as well as in sporophytic aspects. The character that most genera have in common is the inflated alar cells, but in some genera these are absent (*Pterogonidium*) or indistinct (*Meiothecium* pp.). Even more heterogeneous are the sporophytes in this family. Roughly two types of peristomes can be distinguished: the complete peristome with thick, brown exostome teeth, on outer surface transversely striolate in the basal part and on inner side thickened with high transverse lamellae; the endostome in this type is well-developed with a high basal membrane and broad, keeled segments and cilia. This type of peristome is the most common in the family. The second type has narrow, pale exostome teeth, not transversely striolate and hardly thickened at inner side; the endostome can be absent as in *Meiothecium* and *Pterogonidium* or reduced to a low basal membrane with very slender and fragile segments as in *Potamium*. These reduced endostomes could be considered as derivations of the first type, but the very different structure of the exostome teeth suggests a less close relationship with the other genera of the Sematophyllaceae.

[6] Herbarium Division, Department of Plant Ecology and Evolutionary Biology. Heidelberglaan 2, 3584 CS Utrecht, The Netherlands.

LITERATURE

Brotherus, V.F. 1909. Musci in A. Engler & K. Prantl. Die natürlichen Pflanzenfamilien 1 (3), Leipzig. Sematophyllaceae: 1098-1124.

Buck, W.R. 1982. On Meiothecium. Contr. Univ. Michigan Herb. 15: 137-140.

Buck, W.R. 1983. Nomenclatural and taxonomic notes on West Indian Sematophyllaceae. Brittonia 35: 309-311.

Buck, W.R. 1985. A review of Taxithelium in Brazil. Acta Amazon. supl.15: 43-53.

Buck, W.R. 1986. Wijkia in the new world. Hikobia 9: 297-303.

Buck, W.R. 1988. Donnellia resurrected and refound in Florida after 110 years. Bryologist 91: 134-135.

Buck, W.R. 1989. Miscellaneous notes on Antillian mosses 2. Rhaphidostichum in the New World. Moscosoa 5: 189-193.

Buck, W.R. 1990. Contributions to the Moss Flora of the Guianas. Mem. New York Bot. Gard. 64: 184-196.

Buck, W.R. 1991. The basis for familial classification of pleurocarpous mosses. Advances Bryol. 4: 169-185.

Crum, H. 1971. Nomenclatural changes in the Musci. Bryologist 74: 165-174.

Crum, H. 1977. Meiothecium, a new record for North America. Bryologist 80: 188-193.

Dixon, H.N. 1920. Rhaphidostegium caespitosum (Sw.) and its affinities. J. Bot. 58: 81-89.

Dozy, F. & J.H. Molkenboer 1854: Prodromus Florae Bryologicae surinamensis. Haarlem.

Fleischer, M. 1923: Die Musci der Flora von Buitenzorg 4. Sematophyllaceae: 1171-1374.

Florschütz-de Waard, J. 1990. A catalogue of the Bryophytes of the Guianas II, Musci. Trop. Bryol. 3: 89-104.

Florschütz-de Waard, J. 1992. A revision of the genus Potamium. Trop. Bryol. 5: 109-121.

Mitten, W. 1869. Musci Austro-Americani: Sematophylleae. J. Linn. Soc. Bot. 12: 469-497.

Müller, C. 1848. Plantae Kegelianae Surinamenses, Musci frondosi. Linnaea 21: 181-200.

Thériot, I. 1937. Additions a la flore bryologique de la Colombie. Rev. Bryol. Lichénol. 10: 11-18.

KEY TO THE GENERA

1 Alar cells very large and inflated, 70-170 µm long, curved to the insertion. Leaf apex tubulose by incurved margins · · · · · · · · · · · · · · · *1. Acroporium*
 Alar cells, if inflated, not over100 µm long. Leaf apex not tubulose · · · · · · 2

2 Leaf cells papillose · 3
 Leaf cells smooth · 4

3 Leaf cells multi-papillose · *6. Taxithelium*
 Leaf cells uni-papillose · *7. Trichosteleum*

4 Plants bipinnately branched. Branch leaves considerably smaller than stem leaves · *8. Wijkia*
 Plants irregularly to subpinnately branched. Branch and stem leaves little differentiated in size · 5

5 Alar cells in basal row usually inflated · 7
 Alar cells quadrate, in basal row sometimes oval, but not inflated · · · · · · · 6

6 Leaf cells linear, thin walled · *4. Pterogonidium*
 Leaf cells elongate-rhomboidal, incrassate with fusiform lumen · · · · · · · · ·
 · *2. Meiothecium (commutatum)*

7 Leaf margins narrowly reflexed except near apex. Leaf cells incrassate, basal cells towards margins in oblique rows · · · · · · *2. Meiothecium (boryanum)*
 Leaf margins not or only loosely reflexed · 8

8 Leaves oval-ovate or semi-circular, small (to 1.4 mm long) · · · · · · · · · · · · 9
 Leaves oblong or lanceolate (if oval-ovate longer than 1.4 mm) · · · · · · · · · ·
 · 5. Sematophyllum*

9 Leaf cells thin walled or irregularly incrassate. Peristome with pale, non-striolate exostome teeth and inconspicuous endostome with low basal membrane and filiform or rudimentary segments · · · · · · · · · · *3. Potamium*
 Leaf cells regularly incrassate with fusiform lumen. Peristome with strong, brown exostome teeth, striolate in the basal part, and a well-developed endostome with high basal membrane and broad, keeled segments · · · · · ·
 · *5. Sematophyllum (subpinnatum)*

1. **ACROPORIUM** Mitt., J. Linn. Soc. Bot. 10: 182. 1868.
 Lectotype (Buck 1983): A. brevicuspidatum Mitt.

Slender to robust plants with creeping stems; branches erect or ascending, densely foliate. Leaves erect-spreading, ovate-lanceolate, contracted at base, ecostate; leaf cells linear, incrassate, occasionally uni-papillose, alar cells strongly inflated in 1 row, forming a conspicuous, sharply defined group.

Seta smooth or slightly papillose in upper part; capsule suberect to inclined, exothecial cells collenchymatous, operculum long-rostrate; peristome double, exostome teeth transversely striolate with a median furrow on the dorsal plates, endostome segments from a high basal membrane, cilia present.

Distribution: pantropical, only one species in the Guianas.

1. **Acroporium pungens** (Hedw.) Broth., Nat. Pfl. Fam. 11: 436. 1925. – *Hypnum pungens* Hedw., Spec. Musc. 1: 237. 1801. Type: Jamaica, Swartz s.n. (G). – Fig. 149.

Acroporium guianense (Mitt.) Broth., Nat. Pfl. Fam. 11: 436. 1925. – *Sematophyllum guianense* Mitt., J. Linn. Soc. Bot. 12: 478. 1869. Type: Brazil, Caripe prope Pará, Spruce 832 (NY).

Rather robust, bright green to yellowish green, glossy plants, growing in dense tufts. Stems long, creeping or prostrate, defoliate with age, irregularly divided; branches erect, to 2 cm long, often cuspidate at the tip. Stem leaves loosely appressed, branch leaves concave, patent-spreading, sometimes slightly homomallous, ovate-lanceolate, more or less cordate at base, to 2.2 mm long and 0.7 mm wide, apex tubulose by involute margins, denticulate at extreme tip by prorate apical cells, margins minutely serrulate throughout, involute in upper part; leaf cells smooth or uni-papillose at dorsal side, incrassate, linear, 40-70 µm long and 5-7 µm wide, shorter in apex, shorter and wider near insertion, alar cells rectangular to oblong, the outer 4-6 hyaline, conspicuously inflated, 70-170 µm long and 20-40 µm wide.
Synoicous. Perichaetia small, inner perichaetial leaves convolute, about 1 mm long, broad-ovate with a short-acuminate, serrate apex, alar cells little differentiated. Seta 5-10 mm long, rough below the capsule; capsule suberect to inclined, ovoid-cylindric, operculum conic, with a long and slender rostrum (to 0.8 mm long); peristome with the characters of the genus, exostome teeth to 400 µm long, papillose on dorsal side, with high lamellae on ventral side, endostome segments broad, keeled, papillose, cilia single or double, sometimes rudimentary. Calyptra narrowly cylindric.

Distribution: pantropical.

Ecology: Epiphytic on lianas, tree trunks, roots and branchlets, occasionally epiphyllous, also on decaying wood. Occurs in the under-growth and lower canopy of all types of forest, most common in lowland savanna forest and mountain savanna forest.

Selected specimens: Guyana: Upper slopes of Mt. Makarapan, Alt. 800 m, Maas et al. 7480 (U); Waramadan, trail from Kamarang River to Pwipwi Mt, Alt. 800 m, Gradstein 5694 (U). Suriname: Suriname River, Jodensavanne-Mapane creek area, Lindeman 3927 (U); Lely Mts., edge of Plateau 5, Alt. 650 m, Florschütz 4855(U). French Guiana: Crique Kapiri, R.N.2, Bassin de l'Approuage, Alt. 70 m,

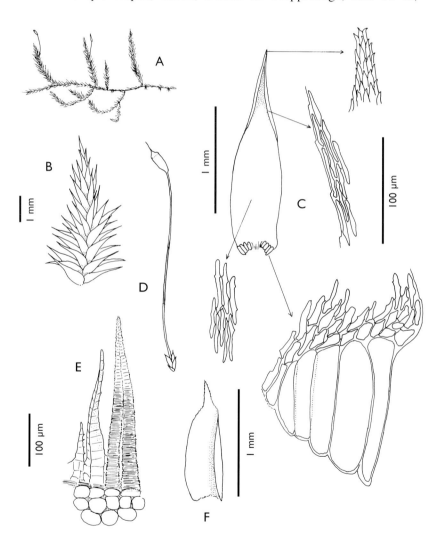

Fig. 149. *Acroporium pungens*: A. habit; B. end of branch; C. branch leaf; D. perichaetium and capsule; E. exostome tooth and part of endostome; F. perichaetial leaf. (Florschütz-de Waard & Zielman 5030).

Cremers 11640 (CAY, U); Mt. de Kaw, 2 km E of Camp Caiman, Alt. 200-300 m, Cornelissen & ter Steege C263 (U).

N o t e : A conspicuous moss, easy to recognize by the erect, cuspidate branches with patent-spreading, pungent leaves. The more or less cordate leaf base with strongly inflated and incurved alar cells distinguish this species from the *Sematophyllum* species of the Guianas.

Acroporium guianense described as *Sematophyllum* (Mitten 1869) is identical to *A. pungens*, only differing in the papillose cells in the upper part of the leaf. It could be considered as a variety of the species, but the presence of papillose leaf cells is extremely variable, even on the same plant. In the Guianas the papillose form is most frequent in the collections of lowland forest, but rare in the collections of higher altitudes. No relations could be noticed with habitat conditions.

2. **MEIOTHECIUM** Mitt., J. Linn. Soc. Bot. 10: 185. 1868.
Type: not indicated.

Slender to medium-sized plants growing in small tufts or loose mats. Stems creeping or prostrate, irregularly branched, branches short, ascending. Leaves more or less concave, ovate-oblong to oblong-lanceolate, costae short and double, indistinct, apex acute or short-acuminate, margin entire, plane or reflexed; leaf cells smooth, elongate-rhomboidal to linear, incrassate, alar cells quadrate or transversely elongate, in the basal row often inflated.
Seta short, smooth; capsule erect or inclined, operculum long-rostrate; peristome single, exostome teeth linear-lanceolate, remote to closely spaced.

D i s t r i b u t i o n : mainly neotropical.

KEY TO THE SPECIES

1 Leaf margins reflexed from below apex to base; basal alar cells distinctly inflated. Exostome teeth remote, papillose · · · · · · · · · · · · *1. M. boryanum*
 Leaf margins plane; alar cells little inflated. Exostome teeth closely spaced, smooth · *2. M. commutatum*

1. **Meiothecium boryanum** (C. Müll.) Mitt., J. Linn. Soc. Bot. 12: 469. 1869. – *Neckera boryana* C. Müll., Syn. 2: 75. 1851. Syntypes: Dominican Republic, Bory s.n. (BM, hb. Besch.); Suriname, Weigelt s.n. (L, NY).   – Fig. 150.

Slender, light green plants, orange-brown with age, growing in loose
mats. Stems prostrate, ascending at the tip, irregularly branched, branches
short, erect, terete. Leaves regularly imbricate when dry, erect-spreading

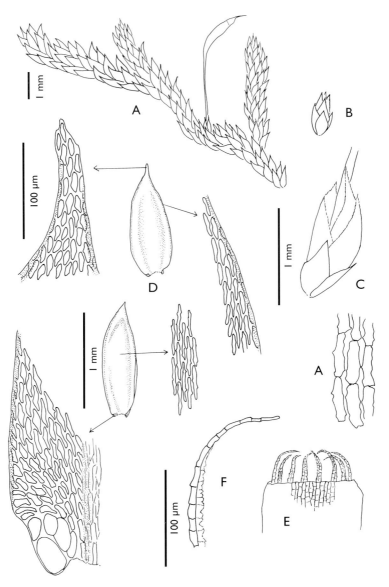

Fig. 150. *Meiothecium boryanum*: A. stem portion with branches and capsule;
B. perigonium; C. perichaetium; D. leaves; E. peristome; F. peristome tooth, side
view; G. exothecial cells, centre of capsule. (Florschütz 1676).

when moist, ovate-oblong, concave except along the border, 0.8-1.4 mm long and 0.3-0.5 mm wide, apex acute, slightly rounded at extreme tip, margin subentire, narrowly reflexed from just below apex to base; leaf cells incrassate, in apex oval or rounded-rhomboidal, 14-20 µm long and 7-9 µm wide, gradually more elongate towards base, at midbase linear, to 50 µm long, strongly incrassate and porose, cells in the marginal part of the leaf base arranged in oblique rows towards the border, alar cells quadrate or transversely elongate, in the basal 1 or 2 rows inflated.

Autoicous. Perigonia small, less than 0.5 mm high, with ovate, acute leaves. Perichaetia about 1 mm high, inner perichaetial leaves ovate-lanceolate with acute or short-acuminate apex, margin serrulate, not reflexed. Seta reddish, smooth, 3-4 mm long; capsule erect or inclined, cylindric, 1-1.5 mm long, with a short neck, exothecial cells elongate, with incrassate and pitted longitudinal walls, operculum slenderly rostrate, about 1 mm high; exostome teeth remote, linear-lanceolate, about 150-200 µm long, pale, papillose on both surfaces. Calyptra scabrous at the tip.

Distribution: West Indies, Central America, tropical South America.

Ecology: Epiphytic in the forest canopy and on stems and branches of solitary trees, occasionally on palm leaf roofs. Not common in the Guianas.

Selected specimens: Guyana: Moraballi creek, Richards 441 (NY, as *Sematophyllum loxense*); Mabura Hill, 180 km SSE of Georgetown, Cornelissen & ter Steege C1073 (U). Suriname: Saramaca River, rock savanna along trail from Paka-Paka to Ebbatop, Florschütz 1586 (U); Area of Kabalebo Dam project, marshforest along road km 212, Florschütz-de Waard & Zielman 5668 (U). French Guiana: Montsinery, Citrus plantation, on Mangifera tree, Aptroot 15513 (U); Saül, 2 km SW of the village, moist lowland forest, Montfoort & Ek 507 (U).

Notes: *Meiothecium boryanum* can easily be confused with *Sematophyllum subpinnatum* which occurs in the same habitat and shows some resemblance in the concave leaves with elliptic, incrassate cells. *M. boryanum*, however, is different in the terete branches (when dry) and the narrower leaves with reflexed margins (even when moist) and lingulate-protruding apex; characteristic is the arrangement of the marginal cells in the leaf base in oblique rows towards the border. The single peristome with widely spaced, papillose teeth is another distinguishing character.

The Weigelt collection from Suriname, syntype of this species, was previously reported as *Pterogonium urceolatum* Schwaegr. by Müller (1848).

2. **Meiothecium commutatum** (C. Müll.) Broth., Nat. Pfl. Fam. 1(3): 1101. 1908. – *Neckera commutata* C. Müll., Bot. Zeit. 15: 385. 1857. – *Meiotheciopsis commutata* (C. Müll.) W.R. Buck, Contr. Univ. Michigan Herb. 15: 137. 1982. – *Donnellia commutata* (C. Müll.) W.R. Buck, Bryologist 91: 134. 1988. Type: Brazil, prope Rio de Janeiro, Beske s.n. (not seen). – Fig. 151.

Meiothecium tenerum Mitt., J. Linn. Soc. Bot. 12: 470. 1869. Type: Brazil, Minas Gerais, Gardner 75 (not seen; see Crum 1977)
Fabronia donnellii Aust., Bot. Gaz. 2: 111. 1877. Type: Florida, St. Augustine, Capt. J.D. Smith (NY).

Slender, light-green plants, growing in thin mats. Stems creeping, sparingly branched, branches short, ascending, curved when dry. Leaves erect, vertical on the creeping stem, at the end of the branches often homomallous, ovate-oblong, to 1.2 mm long and 0.3 mm wide, symmetric to slightly falcate, apex acute or gradually acuminate, margin plane, entire; leaf cells incrassate, elongate-rhomboidal with a fusiform lumen, very homogeneous throughout the leaf, at midleaf 40-70 µm long and 8-10 µm wide, in apex and towards margins slightly shorter, alar cells quadrate in a triangular group, in the basal row coloured and oval, but seldom inflated.
Autoicous. Perigonia small, 0.35 mm high, leaves broadly ovate with short-acute apex. Perichaetia to 0.7 mm high, leaves ovate with acuminate apex. Seta 2-3 mm long, smooth; capsule cylindric, erect or somewhat inclined, exothecial cells irregularly quadrate or rectangular, collenchymatous, operculum long-rostrate; peristome single, exostome teeth closely spaced, about 100 µm long, smooth, in the basal part trabeculate on outer surface. Calyptra narrowly cylindric, smooth.

Distribution: Florida, West Indies, Central America, tropical South America.

Ecology: Apparently rare; only once collected in French Guiana. No habitat details are indicated on the label, but this species is generally growing epiphytic on the lower side of slanted trunks (Buck 1988).

Specimen examined: French Guiana, vicinity of Eaux Claires, ca. 6 km N of Saül, along road to Bélizon, Buck 18695 (NY).

Note: The taxonomic position of this species is not clear. Described as *Neckera*, it was placed in *Meiothecium* sect. *Pterogonidiopsis* by Brotherus (1909) and later transferred to *Meiotheciopsis*, a genus with double peristomes (Buck 1982). *Meiothecium tenerum*, described with a single peristome (Mitten 1869, Brotherus 1909) is identical with *M. commutatum* (Crum 1977). Buck (1988) confirmed the synonymy but left the species in *Meiotheciopsis* (introducing the older name *Donnellia*) defining the peristome as "double with a reduced endostome, which is difficult or impossible to demonstrate". After careful examination of peristomes in the collection of French Guiana no traces of an endostome could be observed, so the place in a genus with double peristomes is at least doubtful. For that reason it is here left in *Meiothecium*, though the differences in peristome characters with *M. boryanum* are considerable.

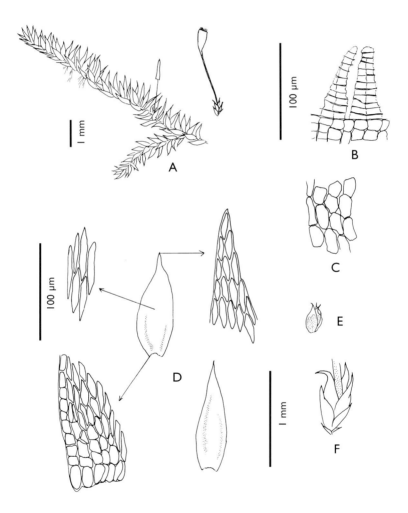

Fig. 151. *Meiothecium commutatum*: A. stem portion with capsule; B. peristome; C. exothecial cells, centre of capsule; D. leaves; E. perigonium; F. perichaetium. (Buck 18695).

3. **POTAMIUM** Mitt., J. Linn. Soc. Bot. 12: 472. 1869.
Lectotype (Florschütz-de Waard 1992): P. vulpinum (Mont.) Mitt.
(Neckera vulpina Mont.). See discussion under Note.

Plants with prostrate stems, irregularly branched, growing in loose mats. Leaves oval or semi-circular; costa short and double, usually indistinct; apex acute, rounded or obtuse; margin subentire. Leaf cells smooth, elongate-rhomboidal at midleaf, rhombic or oval at apex; alar cells quadrate, inflated in the basal row. Autoicous. Seta short, smooth; capsule erect, cylindric, operculum conic, rostrate, exothecial cells rectangular, incrassate along the longitudinal walls, seldom collenchymatous; peristome double, exostome teeth linear or lanceolate, erect when moist, papillose, basal part more or less trabeculate, not or very faintly striolate, endostome consisting of a low basal membrane and filiform or rudimentary segments.

Distribution: tropical S. America.

Note: The genus *Potamium*, created by Mitten (1869), originally consisted of 7 species which showed some superficial resemblance in appearance but which, after thorough study of the peristomes, turned out to mostly belong in the genera *Pterogonidium*, *Meiotheciopsis* and *Sematophyllum*. Only the 2 following species remained in *Potamium*.
The lectotypification of *P. vulpinum* is troubled by the indication of *P. lonchophyllum* as lectotype (Buck 1990); this species, however, was previously transferred to *Sematophyllum* (Florschütz-de Waard 1990). See for a detailed discussion the revision of the genus *Potamium* (Florschütz-de Waard 1992).

KEY TO THE SPECIES

1 Leaves with short-acuminate to broad-acute apex. Exostome teeth linear-lanceolate, gradually tapering, papillose throughout; endostome segments usually rudimentary ···························*1. P. deceptivum*
Leaves with obtuse to rounded-acute apex. Exostome teeth lanceolate, with a broad base, rather quickly narrowed to a slender, papillose apex, often broken (teeth then short and blunt); endostome segments usually persistent ·····
··*2. P. vulpinum*

1. **Potamium deceptivum** Mitt., J. Linn. Soc. Bot. 12: 473. 1869.
Lectotype: Brazil, Rio Negro, Spruce s.n., hb. Mitten 827 (NY, H).
Syntype: ibid, hb. Mitten 826 (NY, H). – Fig. 152.

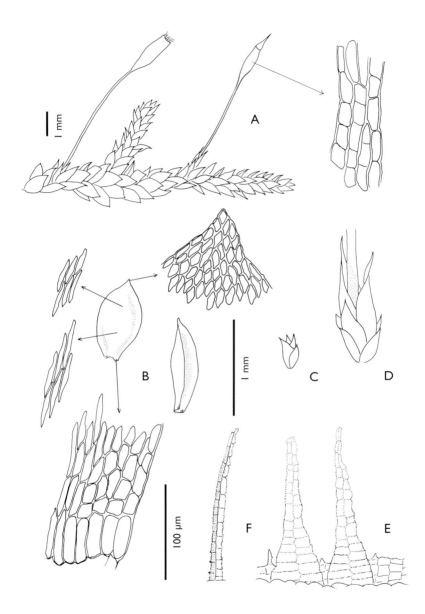

Fig. 152. *Potamium deceptivum*: A. stem portion with capsules and detail of exothecial cells; B. leaves; C. perigonium; D. perichaetium; E. exostome teeth and endostome with rudimentary segments; F. exostome tooth, side view. (Florschütz 2070).

Potamium leucodontaceum (C. Müll.) Broth., Nat. Pfl. 1(3): 1107. 1908. –
Aptychus leucodontaceus C. Müll., Malpighia 10: 517. 1896. Type: Guyana,
Mazaruni River, Quelch s.n. (BM, NY).
Meiothecium negrense Spruce ex Mitt., J. Linn. Soc. Bot. 12: 471. 1869.
Lectotype (Florschütz-de Waard 1992): Brazil, Uanauaca, Spruce 970 (NY).
Maguireella vulpina W.R. Buck, Mem. New York Bot. Gard. 64: 193. 1990,
nom. illeg. Type: Guyana, Upper Demerara/Berbice region, Boom 7148 (NY).

Slender, light green plants, growing in thin mats. Stems creeping, sparingly
branched, branches prostrate or ascending. Leaves erect-spreading, oval-
ovate, to 1.4 mm long and 0.7 mm wide, apex acute to short-acuminate,
margin crenulate, plane; leaf cells thin walled to slightly incrassate,
elongate-rhomboidal, 40-70 μm long and 6-8 μm wide, in apex shorter,
irregularly rhombic or oval, at base more elongated, alar cells rounded-
quadrate, hyaline, to 25 μm wide, in the basal row coloured, oblong, little
inflated, 30-50 μm long and 18-25 μm wide.
Autoicous. Perigonia small, to 0.3 mm high, with ovate, broad-acute
leaves. Perichaetia 1-1.5 mm high with lanceolate, acute leaves. Seta
smooth, 3-6 mm long; capsule erect, cylindric, to 1.5 mm long, operculum
slenderly rostrate; peristome erect-spreading when moist, exostome teeth
pale, papillose, slightly remote at base, about 150 μm long, little or not
trabeculate, endostome reduced to a basal membrane with very fragile,
often rudimentary segments. Calyptra narrowly cylindric.

D i s t r i b u t i o n : Colombia, Venezuela, Guyana, Suriname, Brazil.

E c o l o g y : Epiphytic on branches, trunks and bases of trees in open
vegetation, sometimes temporarily inundated. Not common; only collected
in Guyana and Suriname, not seen from French Guiana.

S e l e c t e d   s p e c i m e n s : Guyana: 5 km W. of Timehri Airport,
Thompson's farm, Maas 2504 (U); Upper Demarara/Berbice region, E of
Linden, Boom 7148 (NY, type of *Maguireella vulpina*). Suriname:
Marowijne River, Bonaparte, Florschütz 528 (U); Corantijn River,
Apoera, Florschütz 2066 (U).

N o t e : This species has much in common with *Sematophyllum subpinnatum*
that occurs in the same habitat. Without capsules the two species,
belonging to different genera, are sometimes hard to distinguish. A useful
character to separate them are the midleaf cells: in *S. subpinnatum*
regularly incrassate with fusiform lumen and seldom over 50 μm long,
in *P. deceptivum* thin walled or variably incrassate and generally longer
(to 70 μm). The most reliable distinction is in the peristomes and it is for-
tunate that both species frequently have sporophytes. The pale, erect
exostome teeth and the fragile, usually rudimentary endostome segments
are the best characters to recognize *P. deceptivum*.

2. **Potamium vulpinum** (Mont.) Mitt., J. Linn. Soc. Bot. 12: 473. 1869. – *Neckera vulpina* Mont., Ann. Sci. Nat. Bot. ser. 2: 203, Pl.4, fig.1. 1835. Type: French Guiana, Sources du Jary, Leprieur s.n., hb. Mont. 7 (PC). – Fig. 153.

Sematophyllum maguireorum W.R. Buck, Mem. New York Bot. Gard. 64: 194. 1990. Type: Brazil, Rio Cauabarí, Maguire et al. 60142 (NY).

Dull green to blackish plants growing in loose mats in temporarily inundated areas. Stems prostrate, defoliate with age, irregularly branched, branches in dry season ascending, julaceous, with light green tips. Leaves loosely appressed when dry, erect-spreading when moist, concave, oval to semi-circular, 0.4-1.4 mm long and 0.4-0.8 mm wide, apex round, obtuse or occasionally rounded-acute, margin crenulate, often narrowly revolute from just below apex nearly to base. Leaf cells thin walled or incrassate, elongate-rhomboidal, at midleaf 25-60 μm long and 5-10 μm wide, towards apex shorter and irregularly rhombic, alar cells rounded-quadrate, to about 20 μm wide, usually forming a well-defined group, in the basal row oblong, more or less inflated, 35-50 μm long and 15-20 μm wide, hyaline.
Autoicous. Perigonia small, about 0.3 mm high, with broad-ovate, acute leaves. Perichaetia to 1 mm high, with ovate, acute leaves. Seta variable in length, 2-9 mm long, smooth; capsule erect, cylindric, operculum slenderly rostrate; exostome teeth strongly incurved when dry, erect when moist, papillose, to 200 μm long, often short and blunt when broken, basal part broad, trabeculate on inner surface, rather quickly narrowed to a fragile, slender upper part, endostome with a low basal membrane and erect, fragile, papillose segments. Calyptra narrowly cylindric.

Distribution: Colombia, Venezuela, the Guianas, Brazil.

Ecology: On stones, roots and overhanging branches in temporarily submerged areas. Rather common in the interior, near the rapids in rivers, but also collected along creeks in the coastal region.

Selected specimens: Guyana: Cuyuni River, Camaria Falls, Richards 847 (NY, U); Kuyuwini River, 150 miles from mouth, A.C. Smith 2565 (U). Suriname: Marowijne River, Toetoe Ondre, Kapasi Island, Geijskes 12 (U); Coppename River, island in Kankantrie Falls, Florschütz & Maas 3159 (U). French Guiana: Inini River, Degrad Fourmi, Cremers 9320 (CAY, U); Crique Gabaret, Bassin de l' Oyapock, Cremers 9998 (CAY, U).

Notes: This species is best recognized by the dark green or blackish colour and by the julaceous branches with short, roundish leaves. In

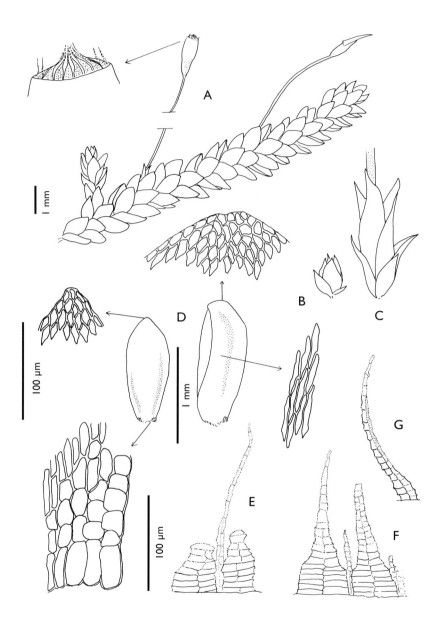

Fig. 153. *Potamium vulpinum* : A. stem portion with capsules; B. perigonium; C. perichaetium; D. leaves; E. peristome with broken exostome teeth; F. peristome with complete exostome teeth; G. exostome tooth, side view. (A-C: Florschütz 1226; D-E: Leprieur s.n., type; F-G: Cremers 5173).

this aspect it resembles the broad-leaved form of *Sematophyllum subpinnatum*, but in the latter species the leaf apex is always short-acuminate or mucronate.

Characteristic for *P. vulpinum* is its preference for a wet, temporarily inundated habitat. It shares this habitat with several semi-aquatic *Sematophyllum* species, e.g. *S. cochleatum, S. pacimoniense* and *S. lonchophyllum*, which all resemble this species in several respects, for instance the leaves with rounded apex and short apical cells. However, in *P. vulpinum* the leaves are smaller and not flaccid. For more detailed differences between these semi-aquatic species see the revision of the genus *Potamium* (Florschütz-de Waard 1992). The separation of these semi-aquatic species remains difficult and sometimes the sporophyte is the only part of the plant to identify the species with certainty. The capsules of *P. vulpinum* with fragile and often broken exostome teeth and slender endostome segments can easily be distinguished from the capsules of the *Sematophyllum* species with strong, transversely striate exostome teeth and broad, keeled endostome segments.

4. **PTEROGONIDIUM** C. Müll., Bull. Herb. Boiss. 5: 209. 1897.
   Type: P. pulchellum (Hook.) C. Müll. (Pterogonium pulchellum Hook.).

Small plants growing in flat mats. Stems elongate, creeping, irregularly branched. Leaves ovate-lanceolate, ecostate; leaf cells linear, smooth, alar cells quadrate.

Seta short, smooth; capsule erect, operculum conic-rostrate, exothecial cells not collenchymatous; peristome single, exostome teeth papillose, not transversely striolate, at inner side not thickened by transverse lamellae.

Distribution: neotropical, only one species in the Guianas.

1. **Pterogonidium pulchellum** (Hook.) C. Müll. in Broth., Nat. Pfl. Fam. 1: 1100. 1909. – *Pterogonium pulchellum* Hook., Musci Exot. 1: 4. 1818. Type: Colombia, Mt. Quindio near el Moral, Humboldt & Bonpland s.n. [not seen. The only Humboldt collections available of this species in (BM) are: Humboldt 57 in hb. Wilson and in hb. Hook. (as nr. 2219); no locality is indicated on the labels]. – Fig. 154.

Potamium casiquiariense Spruce ex Mitt., J. Linn. Soc. Bot. 12: 472. 1869. Syntypes: Venezuela, Rio Casiquiare, Spruce s.n., hb. Mitt. 819 (NY); Brazil, Santarem, Spruce s.n., hb. Mitt. 820 (NY).

Pterogonidium microtheca (C. Müll.) Broth., Nat. Pfl. Fam. 1: 1100. 1909, syn. nov. – *Hypnum microtheca* C. Müll., Linnaea 21: 199. 1848. Type: Suriname, "ad truncus arborum prope Paramaribo", Kegel 518, Aug. 1845 (GOET).

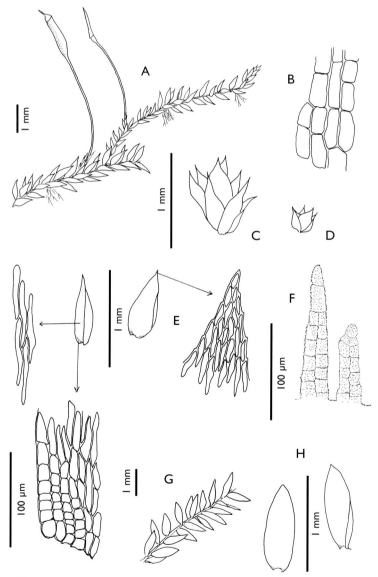

Fig. 154. *Pterogonidium pulchellum*: A. stem portion and branch with capsules; B. exothecial cells of capsule; C. perichaetium; D. perigonium; E. leaves; F. peristome teeth; G. stem portion of submerged form; H. leaves of submerged form. (A-E: Florschütz 4591; F-H: Geijskes s.n.).

Small yellowish green, glossy plants, growing in flat mats. Stems creeping, distantly foliate, irregularly branched; pseudoparaphyllia often present, filiform or subfoliose; branches short, prostrate, with erect-spreading leaves. Leaves ovate or ovate-lanceolate, 0.5-1.2 mm long and 0.2-0.3 mm wide, apex acute to slightly acuminate, sometimes rounded-acute, margin minutely serrulate nearly to base; leaf cells thin walled, linear, at midleaf 60-120 µm long and 5-8 µm wide, in apex shorter, often rhomboidal, alar cells quadrate, arranged in longitudinal rows, forming a distinct triangular group.

Autoicous. Perigonia small, perigonial leaves ovate, about 0.3 mm long. Perichaetia about 1 mm high, perichaetial leaves ovate with acuminate apex. Seta smooth, 2-4 mm long; capsule erect, ovoid or cylindric, about 1 mm long, operculum short-rostrate, exothecial cells rectangular, thin walled or incrassate along the longitudinal walls; peristome with the characters of the genus, exostome teeth blunt, about 100 µm long, with faint median line, strongly papillose. Calyptra narrowly cylindric.

D i s t r i b u t i o n : West Indies, Central America, tropical South America.

E c o l o g y : On trunks and roots of solitary trees, frequently on palm trees, occasionally on charcoal or decaying wood, also on temporarily submerged places. Common in Suriname in the coastal region and on cultivated areas along the rivers. Rarely collected in Guyana and French Guiana.

S e l e c t e d   s p e c i m e n s : Suriname: Paramaribo, Plantation Peperpot on palmtree, Florschütz 1046 (U); Paramaribo, Cultuurtuin, on charcoal in Orchid nursery, Florschütz-de Waard & Zielman 5009 (U); Pikin Rio, near Asidonopo, inundated in wet season (submerged growth form), van Donselaar 1564 (U). French Guiana: Savane Gabrielle, sur un arbre isolé lelong de la Crique Gabrielle, Cremers 3861 (CAY, U).

N o t e s : This slender moss is rather constant in characters and easily recognized by the small, acute leaves with conspicuous, triangular groups of quadrate alar cells. Characteristic is the growth in thin mats on the base of solitary trees, usually densely covered with short sporophytes. When growing in periodically inundated habitats, e.g. at the border of rivers, it produces long, threadlike stems, sparingly branched, with distant, complanate-spreading, broad-acute or rounded leaves. Mitten (1869) described Spruce collections of this form as a species of *Potamium* (*P. casiquiariense*), but the capsules with single peristome and the quadrate alar cells are decisive for *Pterogonidium*.

The species was first recorded for the Guianas as *Hypnum microtheca* (Müller 1848). Under that name the species was also described and illustrated by Dozy & Molkenboer (1854) after a Focke collection. Unfortunately, this collection (L, hb. Miquel 1388) is a mixture of *P. pulchellum* and *Isopterygium subbrevisetum* and the capsules of the latter species have been used for the drawings of the peristome. The gametophytes of both species show much resemblance, but *Isopterygium subbrevisetum* is different in the nearly smooth leaf margins and the lack of a distinct group of alar cells.

5. SEMATOPHYLLUM Mitt., J. Linn. Soc. Bot. 8: 5. 1864.
Type: S. demissum (Wils.) Mitt. (Hypnum demissum Wils.)

Small to medium sized plants growing in flat or rough mats. Stems creeping, irregularly to subpinnately branched, branches horizontal or ascending. Leaves ovate, oblong or lanceolate, costae short and double, indistinct, apex acuminate, acute, rounded and apiculate or obtuse, margin entire, serrulate or serrate; leaf cells elongate-rhomboid, elliptic or linear, thin walled or incrassate; alar cells differentiated, inflated in the basal row. Autoicous or dioicous. Seta elongate, smooth; capsule erect, inclined or pendent, operculum conic-ostrate, exothecial cells rounded quadrate, collenchymatous; peristome double, exostome teeth at inner side with high transverse lamellae, at outer side transversely striolate in lower half, sometimes furrowed, papillose in upper half, endostome with a high basal membrane and keeled, papillose segments and usually well-developed single or double cilia. Calyptra narrowly cylindric.

Distribution: pantropical.

KEY TO THE SPECIES

1 Leaves ovate or oval to semi-circular, 1-2(-3) times as long as wide · · · · · 2
   Leaves oblong, ovate-lanceolate or lanceolate, more than 3 times as long as wide · 4

2 Leaves 0.7-1.2(-1.5) mm long; leaf cells regularly incrassate with fusiform lumina. Plants of sun-exposed habitats · · · · · · · · · · · · · 5. *S. subpinnatum*
   Leaves 1-2 mm long; leaf cells thin walled or irregularly incrassate. Plants of semi-aquatic habitats (if leaves smaller see *Potamium*) · · · · · · · · · · · · · 3

3 Leaves firm, concave (cochleariform) in older leaves; leaf apex rounded and apiculate to short-acuminate (if leaf apex acute see also *Trichosteleum horn-schuchii* var. *subglabrum*). Capsules on a slender seta (6-10 mm long) · *1. S. cochleatum*
   Leaves flaccid, not concave; leaf apex blunt or rounded-acute. Capsules on a short, firm seta (2-3 mm long) · · · · · · · · · · · · · · · · · · *4. S. pacimoniense*

4 Leaves flaccid, oblong to linear with rounded or short-acuminate apex. Plants of semi-aquatic habitats ·····················*3. S. lonchophyllum*
   Leaves firm, concave or flat, lanceolate or oblong with acute-acuminate apex. Plants of dryer habitats ··5

5 Leaves oblong with abruptly acuminate apex and often broadly inflexed margins below apex; leaf cells usually strongly incrassate ········*2. S. galipense*
   Leaves ovate-lanceolate with acute or gradually acuminate apex, margins plane or narrowly reflexed; leaf cells thin walled, occasionally slightly incrassate ··································*6. S. subsimplex*

1. **Sematophyllum cochleatum** (Broth.) Broth., Nat. Pfl. Fam. ed. 2, 11: 433. 1925. – *Rhaphidostegium cochleatum* Broth., Bih. Kongl. Svenska Vetensk. Akad. Handl. 21 Afd. 3(3): 51. 1895. Type: Brazil, S. Paulo, Mosén 2 (H, NY).   – Fig. 155.

Potamium recurvifolium Thér., Rev. Bryol. Lichénol. 10: 17 (fig.7). 1937. Type: Colombia, Rio Yurumangui, Aubert de la Rue s.n. (PC).

Dull-green or brownish plants with bright green innovations, growing in rough mats. Stems elongate, creeping, branches variable in length, prostrate or horizontally spreading from a vertical substrate. Young leaves shriveled when dry, older leaves deeply concave, with reflexed margins when dry, broadly ovate, obovate, oval or nearly orbicular, 1.2-2 mm long and 0.6-1.2 mm wide, apex obtuse or round and mucronate, occasionally rounded-acute, margin subentire, crenulate at apex; leaf cells thin walled, at midleaf elongate-hexagonal, 45-75 µm long and 7-10 µm wide, in apex rhombic or hexagonal, alar cells rounded-quadrate or rectangular, in the basal row coloured, the outer 4-8 conspicuously inflated, to 75 µm long and 25 µm wide.
Autoicous. Perigonia small, leaves broadly ovate, about 0.4 mm high. Perichaetia to 2 mm high, inner leaves convolute, lanceolate with acute apex. Seta smooth, 6-12 mm long; capsule erect or slightly inclined, ovoid-cylindric, with a more or less distinct neck, operculum obliquely rostrate; peristome with the characters of the genus, exostome teeth about 300 µm high, not furrowed, papillose on outer surface, endostome segments slightly papillose at apex, cilia rudimentary.

Distribution: Colombia, Suriname, French Guiana, Brazil.

Ecology: On branches and stones along creeks and rivers, temporarily submerged. Not common but in some areas abundant. Not seen from Guyana.

404

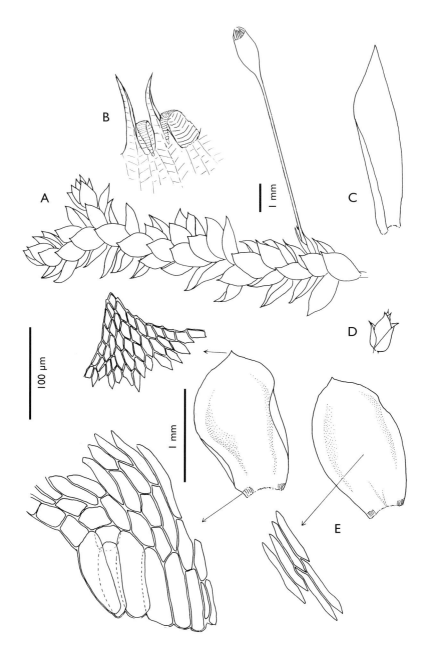

Fig. 155. *Sematophyllum cochleatum*: A. stem portion with capsule; B. peristome from inner side, with incurved exostome teeth; C. perichaetial leaf; D. perigonium; E. leaves (Florschütz 218).

Selected specimens: Suriname: Sara Creek, on branch over water, Florschütz 205 (U); Sipaliwini Savanna, bank of creek bordering forest, on rock, van Donselaar 3724 (U). French Guiana: Haut Tampoc, Saut Awali, sur rocher, Cremers 4745 (CAY, U).

Notes: This species is distinguished from *S. pacimoniense* by the broad, concave leaves which are firm and cochleariform in older parts of the plant. It may resemble the broad-leaved form of *S. subpinnatum*, but the leaves are larger and the leaf cells are thin walled and regularly elongate-hexagonal in the upper leafhalf.
Potamium recurvifolium was described from Colombia and considering the structure of the peristome was placed in section Eu-Potamium (Thériot 1937). All species in this section belong to *Sematophyllum* (Florschütz-de Waard 1992). The leaves of the type are slightly more acuminate than in the type of *S. cochleatum* but this variation can also be observed in the collections from the Guianas.

2. **Sematophyllum galipense** (C. Müll.) Mitt., J. Linn. Soc. Bot. 12: 480. 1869. – *Hypnum galipense* C. Müll., Bot. Zeit. 6: 780. 1848. Type: Venezuela, Galipan, Funck & Schlim 345 (G, BM).

– Fig. 156.

Rhaphidostichum guianense Bartr., Bull. Torrey Bot. Cl. 66: 228. 1939. Type: Guyana, Kanuku Mts., A.C. Smith 3635 (FH, PC, U).

Medium sized, bright green to golden brown, glossy plants growing in rough mats. Stems creeping, partly defoliate, irregularly branched, branches ascending, more or less turgid. Stem leaves small, ovate, often deciduous, branch leaves imbricate, concave, oblong, 1.2-2 mm long and 0.3-0.6 mm wide, at apex rather abruptly narrowed to a short acumen, margin entire or crenulate near apex, usually broadly inflexed below the acumen; leaf cells irregularly incrassate and porose, at midleaf linear, flexuose, 45-75 µm long and 6-9 µm wide, towards apex slightly shorter, flexuose-oblong, at base more incrassate and coloured, alar cells quadrate or oval, often incrassate, in the basal row 3 or 4 strongly inflated, to 75 µm long and 25 µm wide, hyaline.
Autoicous. Perigonia small, budlike. Perichaetial leaves lanceolate, to 2 mm long with acute, serrulate apex. Seta smooth, reddish, to 2 cm long; capsule inclined to horizontal, ovoid, operculum conic, short-rostrate; peristome with the characters of the genus, exostome teeth to 400 mm long, not furrowed, transversely striate in lower 3/4, slightly papillose in upper part, endostome segments finely papillose in upper part, cilia single, well developed.

Distribution: West Indies, Central and tropical South America.

Ecology: On rocks or decaying wood, usually at higher altitudes; not common in the Guianas, not collected in French Guiana.

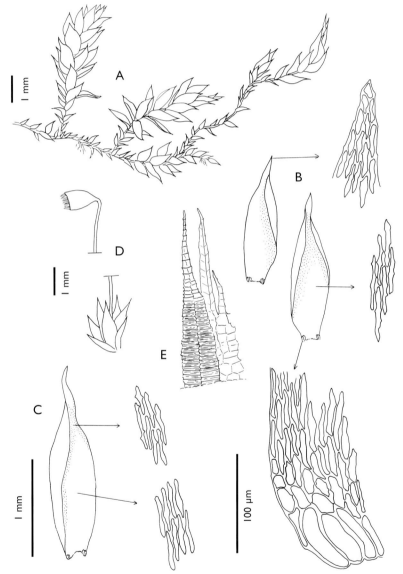

Fig. 156. *Sematophyllum galipense*: A. stem portion with branches; B, C. leaves; D. perichaetium with capsule; E. exostome tooth and part of endostome. (A-B: Florschütz 1477; C-E: Smith 3635, type of *Rhaphidostichum guianense*).

S e l e c t e d   s p e c i m e n s : Guyana: W extremity of Kanuku Mts., Alt. 250 m, A.C. Smith 3119 (PC, NY). Suriname: Van Asch van Wijcks Mts., NW of Ebbatop, Alt. 700 m, Florschütz 1477 (U); Saramaca River, N of Paka-paka, summit of Jan Basi Gado, Alt. 430 m, Florschütz 1650 (U).

N o t e s : *Rhaphidostichum guianense*, described from Guyana is identical with *S. galipense* (Buck 1989). The leaf cells are described as "not incrassate", but in the type collection the cells are distinctly incrassate; the thickening of the cell walls prove to be variable, even in the leaves of one plant.

S. galipense is characterized by the concave, oblong leaves with usually broadly inflexed margins below the short acumen. From *S. subpinnatum* specimens with long, acuminate leaves it can be distinguished by the larger size, the less regular cell pattern and the more conspicuously inflated alar cells.

3. **Sematophyllum lonchophyllum** (Mont.) J. Florsch., Trop. Bryol. 3: 96. 1990. – *Hypnum lonchophyllum* Mont., Syll. Gen. Spec. Pl. Crypt.: 10. 1856. – *Potamium lonchophyllum* (Mont.) Mitt., J. Linn. Soc. Bot. 12: 473. 1869. Type: French Guiana, Cayenne, Leprieur 1378 (PC). – Fig. 157.

Potamium octodiceroides (C. Müll.) Broth., Nat. Pfl. Fam. 1(3): 1107. 1908. – *Ligulina octodiceroides* C. Müll., Hedwigia 40: 84. 1901. Type: Brazil, Restinga de Mauá, Ule 2095 (FH).
Potamium uleanum Broth., Hedwigia 45: 286. 1906. Type: Brazil, Manáus, Ule 255 (FH).

Slender plants growing in loose mats on temporarily submerged substrates. Stems elongate, prostrate or floating, sparingly branched, branches often long, distantly complanate-foliate. Leaves flaccid, oblong-lanceolate to linear, sometimes slightly falcate, 1-3 mm long and 0.3-0.6 mm wide, apex acute or rounded, sometimes short-acuminate, margin serrulate, often bluntly serrate near apex; leaf cells thin walled or incrassate, at midleaf linear, flexuose, 50-100 µm long and 5-9 µm wide, in apex shorter, irregularly rhombic to oblong, alar cells oval or oblong, in the basal row elongated and inflated, to 80 µm long, hyaline, often fragile.

Autoicous. Perigonial leaves broad-ovate, 0.2-0.4 mm long. Perichaetial leaves ovate-lanceolate with an acute, serrate apex, about l.5 mm long. Seta reddish, smooth, about l cm long; capsule inclined, ovoid, operculum long-rostrate; peristome with the characters of the genus, exostome teeth to 250 µm long, more or less furrowed, endostome segments papillose, cilia single.

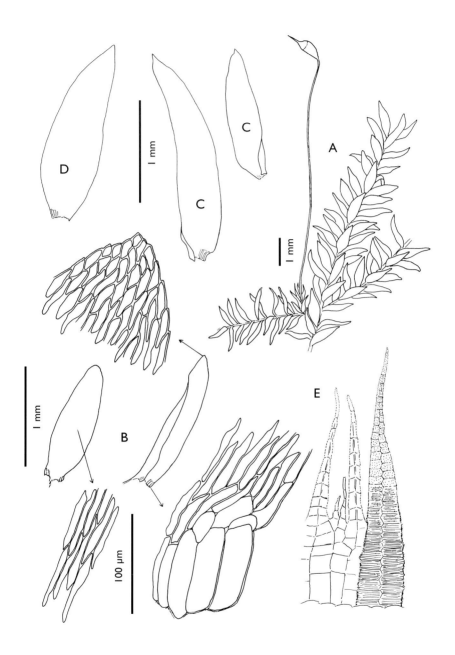

Fig. 157. *Sematophyllum lonchophyllum*: A. stem portion with branches and capsule; B, C, D. leaves; E. exostome tooth with part of endostome. (A-B: Lindeman 5696; C: Leprieur 1378, type; D-E: Lindeman et al. 840).

Distribution: Colombia, Venezuela, Suriname, French Guiana, Brazil.

Ecology: On rocks and decaying wood or terrestrial along rivers and in flooded marsh forest. Not common. Not collected in Guyana, in French Guiana only known from the type locality.

Selected specimens: Suriname: Trail from Wia wia bank to Grote Zwiebelzwamp N of Moengo, Lanjouw & Lindeman 1218 (U); Lely Mts., SW plateau, at margin of pond, Alt. 550-710 m, Lindeman et al. 840 (U); Gran Rio, Gran Dam, in standing water on stone, van Donselaar 1580 (U). French Guiana: Cayenne, Leprieur 1378 (PC, type).

Notes: This species, described as *Potamium* by Mitten (1869), belongs in *Sematophyllum* according to the identical peristome structures (Florschütz-de Waard 1990). The only varying detail is the furrow on the exostome teeth, originating from the thickening of the outer exostome plates as transverse ridges (the transverse striolation). The width of the furrow, depending on the length of the ridges, is a variable character and seems to be of minor taxonomic importance (see also Florschütz-de Waard 1992).
S. lonchophyllum is distinguished from the other *Sematophyllum* species by the long and flaccid leaves with linear leaf cells and fragile alar cells that are elongated but not strongly inflated. The leaf length is rather variable within one collection; on plants with leaves to 2.5 mm long, branches with leaves of hardly 1 mm may occur. The leaf shape is rather constant within one collection, but varies in different collections from lanceolate-linear with an acute apex (as in the type) to oblong with a round apex (as in the type of *P. octodiceroides*). Many intermediate collections make it impossible to distinguish here more than one species.

4. **Sematophyllum pacimoniense** (Mitt.) J. Florsch., Trop. Bryol. 3: 96. 1990. – *Potamium pacimoniense* Mitt., J. Linn. Soc. Bot. 12: 474. 1869. Lectotype (Florschütz-de Waard 1992): Venezuela, Rio Pacimoni, Uaiauca, Spruce s.n., hb. Mitten 829 (NY).   – Fig. 158.

Medium-sized, dull green plants, growing in loose mats in temporarily inundated areas. Stems to 10 cm or more long, floating or prostrate, sparingly branched, branches short. Leaves distant, flaccid, oval-oblong, to 2 mm long and 1 mm wide, apex obtuse or round, occasionally broad-acute, margin entire, crenulate at apex; leaf cells thin walled, at midleaf elongate-hexagonal, 50-90 µm long and 8-15 µm wide, in apex rhombic, alar cells hyaline, thin walled, quadrate or rectangular, the outer 2 or 3 inflated to 100 µm long.

Autoicous. Perigonia small, leaves broad-ovate with short-acuminate apex. Perichaetial leaves about 1 mm long, ovate with rounded-acute apex. Seta short and firm, 2-3 mm long; capsule erect, ovoid with a short neck, operculum obliquely rostrate; peristome with the characters of the

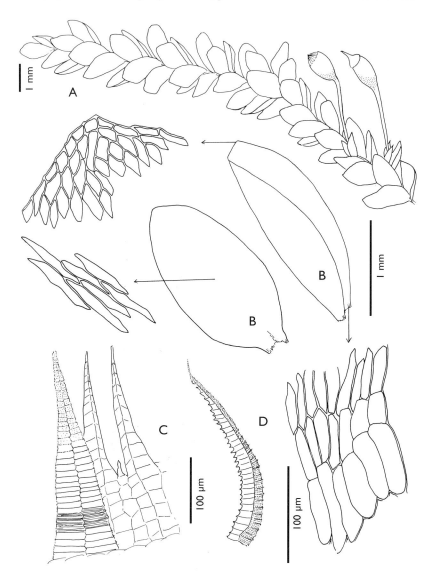

Fig. 158. *Sematophyllum pacimoniense*: A. stem portion with capsules; B. leaves; C. exostome tooth with part of endostome; D. exostome tooth, side view. (Florschütz 4784).

genus, exostome teeth brown, to 350 µm high, not furrowed, endostome segments slightly papillose at apex, cilia rudimentary or absent.

Distribution: Brazil, Venezuela, Suriname.

Ecology: In Suriname only collected on *Montrichardia* stems and overhanging branches along the Casipura creek. Not seen from Guyana or French Guiana.

Specimens examined: Suriname: Casipura creek, on *Montrichardia* stems, just above waterlevel, Florschütz 4784 (U); Casipura creek, on overhanging branches, Moonen s.n. (U)

Notes: Mitten (1869) described *S. pacimoniense* as a species of *Potamium*, but considering the peristome structure it proved to belong in *Sematophyllum* (Florschütz-de Waard 1990, 1992). The species is best recognized by its elongated stems with distant, obtuse leaves and the ovoid capsules on short, firm setae.
Without capsules it is sometimes difficult to distinguish this species from *Potamium vulpinum* that grows in the same submerged habitat. However, the leaves are larger, flaccid and more distantly inserted.

5. **Sematophyllum subpinnatum** (Brid.) Britt., Bryologist 21: 28. l918. – *Leskea subpinnata* Brid., Spec. Musc. 2: 54. l812. Type: Hispaniola, ad arbores, Poiteau s.n., hb. Brid. 747 (B) – Fig. 159.

Sematophyllum caespitosum Mitt., J. Linn. Soc. Bot. 12: 479. l869. – *Leskea caespitosa* Sw. (non Hedw. 1801), Fl. Ind. Occ. 3. 1806, nom. illeg. (see Buck 1983). Type: "Fl. Ind. Occ.", Swartz s.n., hb. Hooker 749 (BM).
Sematophyllum kegelianum (C. Müll.) Mitt., J. Linn. Soc. Bot. 12: 486. 1869, syn. nov. – *Leskea kegeliana* C. Müll., Linnaea 21: 198. 1848. Type: Suriname, Poelepantje, Kegel 513 (GOET).
Rhaphidostegium grammicarpum (C. Müll.) Par., Ind. Bryol.: 1095. 1898, syn. nov. – *Aptychus grammicarpus* C. Müll., Malpighia 10: 512. 1896. Type: Guyana, Marshall falls, Quelch s.n., hb. Lev. 1286 (BM).
Sematophyllum apaloblastum (C. Müll.) W.R. Buck, Brittonia 35: 330. 1983, syn. nov. – *Aptychus apaloblastus* C. Müll., Bull. Herb. Boissier 5: 212. 1897. Type: Guatemala, Pansamalá, v. Türckheim s.n., hb. Broth. nr.16 (H).

Slender to medium sized, usually light green plants growing in loose mats. Stems creeping, irregularly branched, branches ascending, often curved when dry. Leaves loosely imbricate or patent-spreading, on curved branches homomallous, ovate, oval or obovate, sometimes suborbicular, concave in central part, 0.7-1.2(-1.5) mm long and 0.3-0.9 mm

412

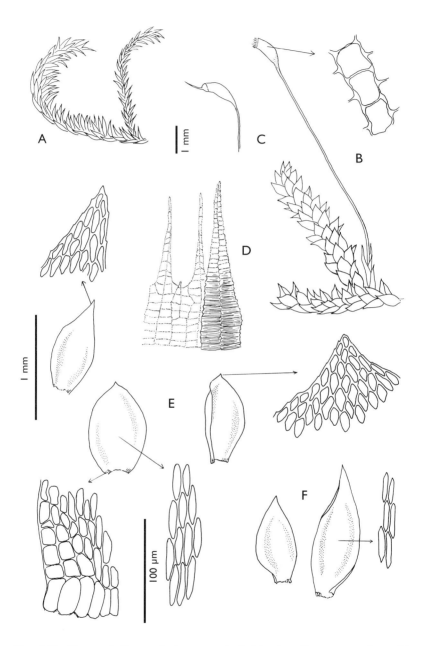

Fig. 159. *Sematophyllum subpinnatum*: A. habit, dry; B. stem portion with branch and capsule; C. capsule with operculum; D. exostome tooth and part of endostome; E, F: leaves. (A-E: Florschütz 4563; F: Türckheim, hb. Broth.16, type of *S. apaloblastum*).

wide, apex acute, short-acuminate or rounded and mucronate, margin subentire or minutely serrulate; leaf cells regularly incrassate with fusiform lumina, at midleaf elongate-rhomboidal, 30-55 µm long and 7-8.5 µm wide, apical cells rhomboidal or oval, 17-35 µm long and 7-8.5 µm wide, basal cells more elongate, porose, alar cells quadrate to oval in a conspicuous triangular group, in the basal row oblong, the outer 3-8 inflated, 30-55 µm long and 18-30 µm wide, sometimes transversely divided.

Autoicous. Perigonial leaves about 0.3 mm long, broad-ovate to semi-circular with short-acuminate apex. Perichaetial leaves to 1.7 mm long, oblong with acute apex, margin subentire. Seta smooth, 6-14 mm long; capsule erect or inclined, ovoid-cylindric, often slightly arcuate, operculum obliquely long-rostrate, exothecial cells rounded-quadrate and collenchymatous, but occasionally more elongate and hardly collenchymatous; peristome with the characters of the genus, exostome not furrowed, about 300 µm long, endostome segments papillose, cilia single, often rudimentary.

Distribution: West Indies, Central and tropical South America, tropical Africa.

Ecology: Epiphytic, often on solitary trees, also on decaying wood, palmleaf roofs and occasionally on stones. A common species in sun-exposed habitats: cultivated areas, riversides, xeromorphic scrub and forest canopy.

Selected specimens: Guyana: Timehri, Dakara creek, Thomsons farm, Gradstein 4723 (U); Kamarang, along trail to Maramadam, Aptroot 17062 (U). Suriname: Coppename River, Raleigh Falls, on palmleaf roof, Florschütz 4563 (U); Area of Kabalebo Dam project, granitic rock slab along road to Amatopo km 117, Florschütz-de Waard & Zielman 5616 (U). French Guiana: Mt. de l'Inini, extrémité Est, foret sur crête, Alt. 700 m, Cremers 9251 (CAY, U); Mt. de Kaw, 2 km N of Camp Caiman, Alt. 200-300 m, Cornelissen & ter Steege C259 (U).

Notes: This variable species preferably grows in exposed habitats, where it is subject to desiccation. Its typical expression in this habitat is characterized by curved branches with homomallous, broad leaves. In more humid habitats, e.g. in open areas of mesophytic rainforest or in the lower part of the canopy, the branches are erect and the leaves are patent-spreading, usually smaller and more slenderly ovate.

The variation in leaf shape runs from suborbicular with a rounded, mucronate apex to slenderly ovate with an acute apex. Between these extremes all intermediates occur, so there is no base to distinguish varieties. This variation has led to the description of many new species, which for the greater part have since been synonymized with *S. caespitosum* by Dixon (1920). This name is later replaced by *S. subpinnatum* because the basionym *Leskea caespitosa* Sw. is an illegitimate homonym of *L. caespitosa* Hedw., which represents a species of *Acroporium* (Buck 1983).

A few collections from higher altitudes (Ebbatop, Bakhuis Mts.) have more elongated leaves, to 1.5 mm long, with long-acute apices. These closely resemble *S. apaloblastum.*, a species described from Guatemala. The Türckheim collections of the latter species, including the type (H), show the same variation in leaf shape as *S. subpinnatum*, so it can be considered as synonymous.

The regular cell pattern of incrassate, fusiform cells, arranged in diverging rows is a typical character for *S. subpinnatum*. In this respect it could only be confused with *Meiothecium commutatum*, but it differs in the broader, concave leaves and the more inflated alar cells. Also *Trichosteleum hornschuchii* var. *subglabrum* with sometimes very indistinct papillae could be mistaken for *S. subpinnatum*, but in that species the midleaf cells are narrower and linear. For differences with *S. galipense* and the *Potamium* species see under those species.

6. **Sematophyllum subsimplex** (Hedw.) Mitt., J. Linn. Soc. Bot. 12: 494. 1869. – *Hypnum subsimplex* Hedw., Spec. Musc.: 270 (Tab. 69). 1801. Type: India occidentalis, Swartz s.n. (this collection could not be located).   – Fig. 160.

Hypnum richardii Schwaegr., Spec. Musc. Suppl. 1(2): 204, (Tab. 93). 1816. Type: French Guiana, Richard s.n. (G).

Slender, glossy plants, growing in flat or rough mats. Stems creeping, irregularly to subpinnately branched, branches short and complanate-spreading or more elongate and ascending, occasionally abruptly ending and forming clusters of propagulae. Stem leaves complanate or more or less homomallous, branch leaves complanate or erecto-patent; leaves ovate-lanceolate, (0.4)0.6-1.4(1.7) mm long and 0.2-0.4 mm wide, apex acute or gradually acuminate, margin minutely serrulate throughout, sometimes serrate towards apex, often narrowly reflexed in upper half; leaf cells thin walled or more or less incrassate, linear, often flexuose, at midleaf 60-120 μm long and 3.5-7 μm wide, shorter and sometimes prorate in apex, shorter and more incrassate at leaf base, alar cells subquadrate in a small group, in the basal row 3-4 conspicuously inflated, to 80 μm long and 30 μm wide.

Autoicous. Perigonial leaves ovate, about 0.3 mm long, with short-acuminate, serrulate apex. Perichaetial leaves lanceolate, to 1.7 mm long

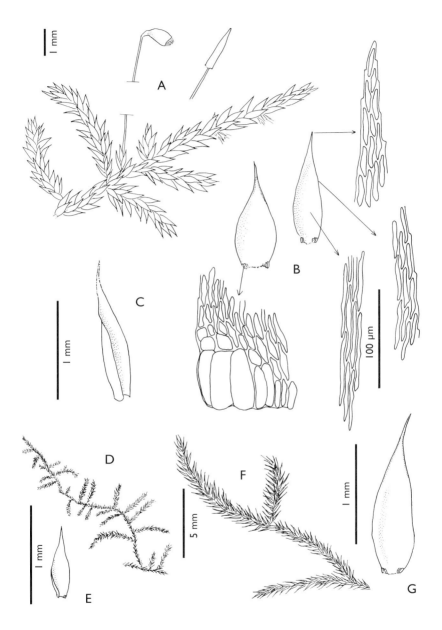

Fig. 160. *Sematophyllum subsimplex*: A. stem portion with branches and capsules; B. leaves; C. inner perichaetial leaf; D. habit of "wiry" form; E. leaf of "wiry" form; F. habit of elongated form; G. leaf of elongated form. (A-C: Florschütz & Zielman 5472; D-E: Florschütz de Waard & Zielman 5774; F-G: Maas 3379).

with gradually acuminate, serrulate apex. Seta smooth, reddish brown, variable in length, 0.5-2.5 cm long; capsule horizontal to pendent, ovoid, operculum obliquely rostrate; peristome with the characters of the genus, exostome teeth to 400 μm long, not furrowed, endostome segments papillose, cilia single or double, well developed.

D i s t r i b u t i o n : West Indies, Central America, tropical South America, tropical Africa.

E c o l o g y : A common moss on decaying wood and tree bases in the lower story of all types of rainforest, also epiphytic on branches of the lower canopy. Also collected in xeromorphic scrub vegetation, not collected in the coastal region.

S e l e c t e d   s p e c i m e n s : Guyana: NW slopes of Kanuku Mts., in drainage of Moku-moku creek, Alt. 150-400m, A.C. Smith 3415 (NY, U); Mabura Hill, near Demarara River, on lower canopy branches, Cornelissen & ter Steege 960 (U). Suriname: Zanderij, savanna forest, Florschütz 4777 (U); Nickerie River, 2 km N of Kamisa Falls, shrub savanna, Maas 3379 (U) (elongated growth form); Area of Kabalebo Dam-project, road to Amatopo, km 212.5, near small rapid in creek in dark rainforest, Florschütz-de Waard & Zielman 5774 (U) (wiry growth form). French Guiana: Mt. Tortue, Bassin de la Comté, Alt. 200m, Cremers 10093 (CAY, U); Mt. de Kaw, 3 km N of Camp Caiman, Alt. 200-300m, Cornelissen & ter Steege 319A (U) (wiry growthform with propagulae).

N o t e s : This species is usually easily recognized in the closely branched growth form with erect-spreading leaves, showing the dark-brown branches. But other expressions occur and the variability of nearly all characters make it difficult to determine the species outline. The leaf shape may vary from ovate to slenderly lanceolate even along one stem; the leaf apex is usually acute and serrulate, but seems often acuminate and entire by the narrowly reflexed margins; the leaf cells are thin walled or incrassate and very variable in length; the inflated alar cells can vary in length from 35-80 μm on the same plant; the seta length ranges from 0.5-2.5 cm but is in one collection remarkably constant. The most confusing variation however is in the size and the aspect of the plants. Two extreme expressions can be distinguished:
1. The small, "wiry" growth form, subpinnately branched with short, horizontal branches, and tiny, complanate-spreading leaves (0.4-0.6 mm long) with acute, serrulate apex and rather short leaf cells.
2. The elongated growth form, sparingly divided, with long, terete branches and long-acuminate leaves (to 1.5 mm long) with narrowly reflexed margins and narrow, linear cells.

The wiry growth form is usually collected in moist places, e.g. on tree foots and logs on the forest floor. The elongated form grows in more open vegetations like savanna forest or xeromorphic scrub. The correlation with different habitats implies that ecological conditions are responsible for this great variation. In the numerous collections examined all kinds of intermediate forms could be observed, so there is no base for distinguishing varieties.

Hypnum richardii was already mentioned as "forma minor" of *S. subsimplex* by Mitten (1869). The type collection from French Guiana (G) is a representative of the "wiry" form. The elongated growth form could be confused with *S. galipense* but in *S. subsimplex* the leaves are less concave and more gradually acuminate with serrulate or narrowly reflexed margins near apex (subentire and broadly inflexed in *S. galipense*).

6. **TAXITHELIUM** Spruce ex Mitt., J. Linn. Soc. Bot. 12: 496. 1869. (Taxithelium Spruce, Cat. Musc.: 14. 1867 (nom. nud.).
Type: T. planum (Brid.) Mitt. (Hypnum planum Brid.).

Hypnum subsect. *Sigmatella* C. Müll., Syn.2: 263. 1851. – *Sigmatella* (C. Müll.) C. Müll., Bot. Jahrb. 23: 328. 1896, hom. illeg.

Slender plants with creeping stems, irregularly to pinnately branched, usually complanate. Leaves ovate or lanceolate, apex obtuse, acute or acuminate, margin subentire, serrulate or dentate, costa short and double or absent; leaf cells linear, seriate-papillose at dorsal surface, alar cells, if differentiated, quadrate or oblong, often partly inflated.
Autoicous. Seta elongate, smooth; capsule inclined to pendent, ovoid-cylindric, operculum conic, short-rostrate; peristome double, exostome teeth with faint median zig-zag line, not furrowed, transversely striate in lower part, papillose in upper part, at inner surface with high transverse lamellae, endostome with high basal membrane and broad, keeled segments and slender cilia. Calyptra cucullate.

D i s t r i b u t i o n : pantropical.

KEY TO THE SPECIES

1 Leaves oval or ovate, with a distinct group of alar cells, quadrate or oblong, sometimes partly inflated. Midleaf cells with 4-10 papillae · · · · · · · · · · · 2
  Leaves lanceolate, with little differentiated alar cells. Midleaf cells with 3-6 papillae · *3. T. pluripunctatum*

2   Branch leaves ovate with acute-acuminate apex. Alar cells limited (seldom more than 12), basal cells usually inflated · · · · · · · · · · · · · · · *2. T. planum*
    Branch leaves oval with broad-acute to round apex. Alar cells quadrate, numerous (10-25) arranged in longitudinal rows, seldom inflated · · · · · · ·
    · *1. T. concavum*

1. **Taxithelium concavum** (Hook.) Spruce, Cat. Musc.: 14. 1867 (comb. inval.) – *Hypnum concavum* Hook. in Kunth, Syn. Pl. Aequinoct.1: 63. 1822. Type: Venezuela, Rio Negro, San Carlos, Humboldt 34, hb. Hook. 3176 (BM).   – Fig. 161.

Taxithelium quelchii (C. Müll.) Par., Ind. Bryol.: 1262. 1898. – *Sigmatella quelchii* C. Müll., Malpighia 10: 519. 1896. Type: Guyana, Marshall Falls, Quelch s.n., hb. Levier 1288 (BM).

Dark green to bronze-coloured plants with light-green innovations, growing in rather dense mats. Stems elongate, irregularly branched, branches to 1 cm long, complanate or terete, often curved when dry. Stem leaves complanate, ovate, with acute, often homomallous apex; branch leaves very concave, oval or cymbiform, 0.6-1.2 mm long and 0.3-0.5 mm wide, apex broad-acute, blunt or round (ventral leaves often acute), margin crenulate or finely denticulate at apex; midleaf cells thin walled or incrassate with a row of 4-8 small papillae, linear, 40-70 µm long and 3-5 µm wide, towards apex shorter, rhomboidal, towards base shorter and wider, incrassate, alar cells arranged in longitudinal rows, forming a conspicuous, sometimes auriculate-inflated group of 10-25 rounded-quadrate cells, in the basal row rectangular or slightly inflated.
Autoicous. Perigonia small, about 0.5 mm high, leaves broad-ovate with short-acuminate apex, papillose. Perichaetia to 2 mm high, leaves lanceolate, slenderly acuminate, serrulate at apex, papillose. Seta smooth, 1-2 mm long; capsule ovoid, curved, erect or inclined, exothecial cells quadrate or hexagonal, little or not collenchymatous, operculum conic, mamillate or bluntly rostellate; peristome with the characters of the genus, exostome teeth about 400 µm long, endostome segments minutely papillose.

D i s t r i b u t i o n : Amazon basin; probably more wide-spread, but usually recorded as *T. planum.*

E c o l o g y : On submerged rocks and soil, also on decaying wood and on branches overhanging the water. Rather common near rapids in the rivers, occasionally on temporarily flooded forest floor.

Selected specimens: Guyana: Basin of Essequibo River, near mouth of Onoro creek, Smith 2655 (NY, U); Mazaruni River, Kartabo Point, Richards 850 (BM). Suriname: Tibiti River, near rapid, van Looy 21 (U); Corantijn River, Frederik Willem IV falls, Florschütz-de Waard

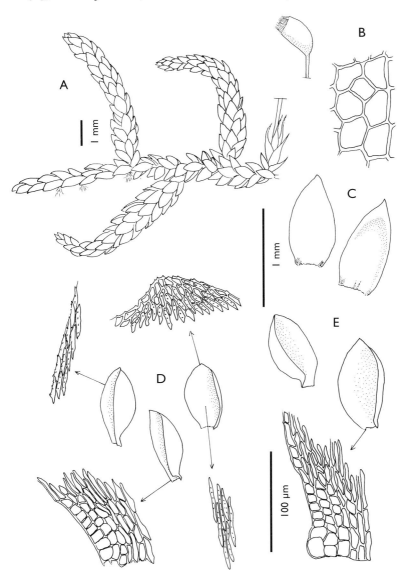

Fig. 161. *Taxithelium concavum* : A. stem portion with branches and capsule; B. exothecial cells of capsule; C. stem leaves; D, E. branch leaves (A-B, E: Florschütz-de Waard & Zielman 5730; C-D: Florschütz-de Waard & Zielman 5183).

& Zielman 5730 (U). French Guiana: Tampoc River, Saut Koumakou soula, Cremers 4723 (CAY, U); Crique Pain de Sucre (confl. of Comté River), on boulder in water, Florschütz-de Waard 5961 (U).

Note: This species, described from the Rio Negro by Hooker (1822), was placed under synonymy of *T. planum* (Mitten 1869). Buck (1985) also considered this species, together with *T. quelchii* from Guyana, as adaptational forms to periodically inundated habitats on rock and soil. However, *T. planum* also sometimes occurs in this habitat without loosing its typical features. *T. concavum* is very constant in the two main differentiating characters: the blunt, cymbiform leaves and the conspicuous group of quadrate cells in the alar region. No transitional forms have been seen in the many collections examined, so at least in the Guianas the two species are well separated. Also the sporophyte is different in the more erect position of the capsule. Other characters are more variable and might be influenced by habitat conditions, for instance the often incrassate cell walls, the less conspicuous papillae and the inconspicuous denticulation of the leaf margins.

2. **Taxithelium planum** (Brid.) Mitt., J. Linn. Soc. Bot. 12: 496. 1869. – *Hypnum planum* Brid., Musc. Recent. Suppl. 2: 97. 1812. Type: Ins. Hispaniola., Poiteau s.n., hb. Brid. 819 (B), hb. Meyer 2 (GOET), isotypes. – Fig. 162.

Pale green to dull green plants growing in often extensive, flat mats. Stems long, creeping, irregularly to pinnately branched; branches short, horizontal or ascending, complanate or terete. Stem leaves complanate, often homomallous, broader at base and more acuminate than branch leaves; branch leaves complanate or loosely imbricate, oval or ovate, concave, contracted at base, 0.5-1.2 mm long and 0.25-0.5 mm broad; apex acute or slightly acuminate; margin serrulate to sharply denticulate, teeth often bifid. Leaf cells thin walled, with seriately arranged, small papillae (1 or 2 on apical cells to 10 on midleafcells, often geminate), midleaf cells linear, 35-85 μm long and 3.5-5 μm wide, apical cells shorter, rhomboidal, basal cells smooth, shorter, wider and incrassate; alar cells forming a small group of hyaline cells, quadrate or irregular, in the basal row oblong and partly inflated, to 40 μm long and 17 μm wide.
Autoicous. Perigonia small, leaves broadly ovate, short-acuminate, papillose. Perichaetia 1.5-2 mm high, leaves lanceolate with slenderly acuminate, flexuose apex, dentate, papillose. Seta smooth, 0.8-2.5 cm long; capsule ovoid, curved, horizontal to pendent, operculum conic, bluntly rostellate, exothecial cells rounded quadrate, variously collenchymatous; peristome with the characters of the genus, exostome teeth to 450 μm high, endostome segments finely papillose.

Ecology: On decaying wood, bark of living trees or rocks. A very common moss in all vegetation types: in the understory of mesophytic rainforest, humid marsh forest, dry savanna forest, also in xeromorphic scrub and cultivated areas; not collected in the forest canopy.

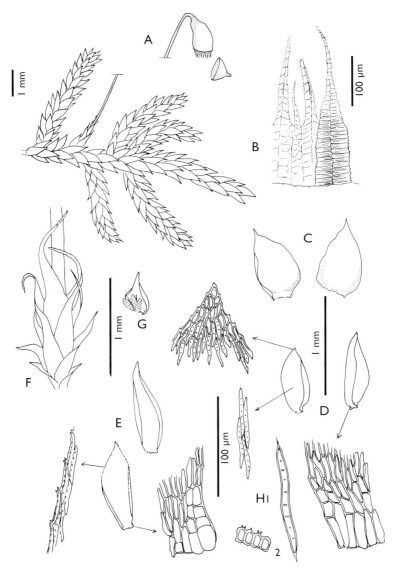

Fig. 162. *Taxithelium planum*: A. stem portion with branches and capsule; B. exostome tooth and part of endostome; C. stem leaves; D, E. branch leaves; F. perichaetium; G. perigonium; H. geminate papillae on leaf cells. (A-D, F-H: Florschütz-de Waard & Zielman 5658; E: Gradstein 4926).

Distribution: Florida, West Indies, Central and tropical South America, tropical Africa.

Selected specimens: Guyana: Kanuku Mts., Rupununi River-Puwib River, Jansen-Jacobs et al. 225 (U); Mazaruni Distr., Jawalla at confluence of Kukui R. and Mazaruni R., Alt. 500 m, Gradstein 4926 (U). Suriname: Paramaribo, Cultuurtuin, Florschütz 578, (terete form)(U); Palaime creek, trib. Sipaliwini R, Wessels Boer 842, (almost smooth specimen)(U). French Guiana: Saül, Sentier Limonade 2 km SW of the village, Alt 180-210 m, Montfoort & Ek 799 (U); Iles du Salut, Ile royale, Cremers 8638 (CAY, U).

Notes: This species is easily recognized by the seriate papillose leaf cells in combination with the leaf shape. The distinctness of the papillae is variable; in a few collections the leaves are hardly papillose, but even then the species can be identified by the oval, concave leaves and the typical denticulate margins with outward-directing, double teeth.
The strongly complanate growth form as indicated by the species name is not the only expression of this species. In sun-exposed areas the branches can be terete with closely imbricate leaves. The leaf apex is always acute or slightly acuminate and the alar cells form a small group with few (usually less than 12) irregular or quadrate cells and with elongated and partly inflated cells in the basal row. In these characters *T. planum* is constant and well-separated from *T. concavum*, even in the specimens that were collected from occasionally flooded rocks, the typical habitat for the latter species.

3. **Taxithelium pluripunctatum** (Ren. & Card.) W.R. Buck, Moscosoa 2: 60. 1983. – *Trichosteleum pluripunctatum* Ren. & Card., Bull. Soc. Roy. Bot. Belgique 29: 184. 1890. Type: Martinique, St. Marie, Bordaz s.n. (NY). – Fig. 163.

Taxithelium patulifolium Thér., Ann. Bryol. 7: 160. 1934. Type: French Guiana, St. Jean du Maroni, collector not indicated (1895), hb. Thériot (PC).

Slender, light green plants, growing in small patches or open mats. Stems creeping, subpinnately branched with short, complanate branches or irregularly branched with more elongated branches. Stem leaves patent-spreading to complanate, ovate-lanceolate, slenderly acuminate; branch leaves complanate-spreading, usually shorter acuminate, subsymmetric to falcate, to 1.2 mm long and 0.3 mm wide, ventral leaves narrower, symmetric. Median leaf cells linear, 60-100 µm long and about 6 µm wide with 3-6 blunt papillae at dorsal side, apical cells with 0-3 papillae, basal cells shorter, smooth, along insertion oval-oblong; alar cells hardly differentiated.

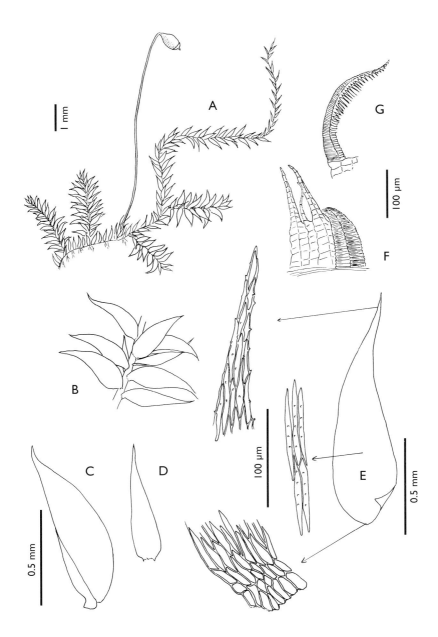

Fig. 163. *Taxithelium pluripunctatum*: A. stem portion with branches and capsule; B. part of branch; C. lateral branch leaf; D. ventral branch leaf; E. stem leaf; F. exostome tooth and part of endostome; G. exostome tooth, side view (A-E: Florschütz 1780; F-G: Tjon Lim Sang & van der Wiel 52).

Autoicous. Perigonia small, to 0.5 mm high, leaves broad-ovate, abruptly acuminate. Perichaetial leaves lanceolate, long-acuminate, to 1 mm long, entire or slightly serrulate. Seta reddish, 0.5-1 mm long; capsule inclined, curved-cylindric, operculum conic-rostrate, exothecial cells rounded-quadrate or more elongate with thickened longitudinal walls; peristome with the characters of the genus, exostome teeth to 350 µm high, endostome segments finely papillose.

Distribution: West Indies, the Guianas, Brazil.

Ecology: Epiphytic on tree bases, roots and branchlets in the understory of humid forests, also on stones and decaying wood in moist areas. In the Guianas not common or probably overlooked.

Selected specimens: Guyana: Santa Mission, swamp forest near the village, Florschütz-de Waard 6135 (U); Mabura region, Ekuk comp., Holder Falls, Ek & Maas 964 A (U). Suriname: Brownsberg, trail to Irene Falls, Alt. ca. 400 m, Tjon Lim Sang & van der Wiel 52 (U); Area of Kabalebo Dam project, marsh forest E of road to Amatopo km 80, Florschütz-de Waard & Zielman 5573 (U). French Guiana: 20 km E-NE de Saül, vallée humide 5 km N de Saut Maïs, de Granville 5906 (CAY, U); Mt. de Cacao, 45 km S of Cayenne, Alt. 150 m, Aptroot 15567 (U).

Note: This tiny moss is usually easily recognized by its pinnate-branched habit with complanate-spreading leaves. In the less regularly divided specimens the more elongated and not complanate branches are identical to the stems.

7. **TRICHOSTELEUM** Mitt., J. Linn. Soc. Bot. 10: 181. 1868.
Lectotype (Fleischer 1923): T. fissum Mitt.

Slender to medium sized plants with creeping stems, irregularly branched. Leaves erect-spreading to slightly complanate, ovate, oblong or lanceolate, apex acute or acuminate, margin subentire to serrate, costa short and double, often indistinct or lacking; leaf cells uni-papillose on dorsal leaf side, alar cells inflated in the basal row.
Seta slender, smooth or papillose in upper part; capsule horizontal to pendent, operculum long-rostrate, exothecial cells collenchymatous; peristome teeth with or without median furrow at dorsal side, transversely striolate in lower part and finely papillose in upper part, ventral plates narrower than dorsal plates, with strong and high transverse lamellae, endostome segments from a high basal membrane, keeled, cilia single or double. Calyptra cucullate, narrowly cylindric.

or double. Calyptra cucullate, narrowly cylindric.

N o t e : The character of a median furrow, present or not at the dorsal side of the exostome teeth, is often considered to be of taxonomical importance. It seems to be dependent on the length of the transverse thickenings of the dorsal plates, seen as a striolation. In *Trichosteleum* this character is variable, even sometimes in one species.

D i s t r i b u t i o n : pantropical.

KEY TO THE SPECIES

1 Leaves with long-acuminate, often flexuose apex · · · · · · · · · · · · · · · · · · 2
 Leaves acute or gradually acuminate · 3

2 Papillae conspicuous. Leaf margin serrate in upper half. Perichaetia small, to
 1.5 mm high · *4. T. papillosum*
 Papillae indistinct, only visible in side-view on the concave part of the leaf.
 Leaf margin weakly serrulate in upper half. Perichaetia conspicuous, to 3
 mm high · *1. T. bolivarense*

3 Leaves lanceolate with slenderly acute or gradually acuminate apex · · · · · ·
 · *3. T. intricatum*
 Leaves ovate or oval-elliptic, with acute, sometimes rounded apex · · · · · · 4

4 Leaves erect-spreading or complanate, flexuose when dry, ovate, 0.7-1.2 mm
 long. Median leaf cells thin walled, elongate-rhomboidal to linear with
 distinct papillae · · · · · · · · · · · · · · · *2a. T. hornschuchii* var. *hornschuchii*
 Leaves imbricate, concave, oval-elliptic, 1-2 mm long. Median leaf cells
 incrassate, flexuose-linear; papillae indistinct or absent, sometimes only
 visible in young leaves · · · · · · · · · · · *2b. T. hornschuchii* var. *subglabrum*

1. **Trichosteleum bolivarense** Robins., Acta Bot. Venez. 1: 78. 1965. Type: Venezuela, Cerro Venamo, Steyermark & Dunsterville 92253 (US); paratype: Steyermark 93551 (NY).   – Fig 164.

Rather robust, light-green, glossy plants, growing in rough mats. Stems elongate, creeping or prostrate, irregularly to pinnately branched, branches short and ascending or sometimes elongate and pendulous. Leaves loosely imbricate, concave, elliptic or elliptic-lanceolate, 1.5-2 mm long and 0.4-0.6 mm wide, apex slenderly acuminate, often flexuose, margin weakly serrulate, partly reflexed in upper part; midleaf cells linear, incrassate, porose, 70-100 µm long and 5-7 µm wide, slightly shorter

426

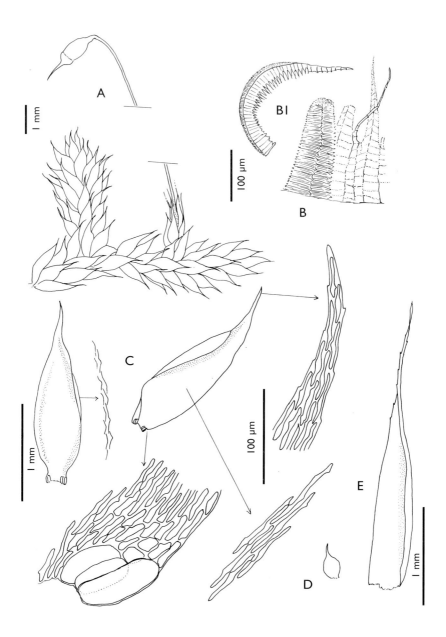

Fig. 164. *Trichosteleum bolivarense*: A. stem portion with branch and capsule;
B. part of peristome with incurved exostome tooth; B1. exostome tooth in side
view; C. branch leaves; D. perigonial leaf; E. perichaetial leaf. (A-C: Florschütz
4798; D-E: Florschütz & Maas 2993).

short and strongly incrassate, coloured, alar cells rounded-quadrate or oblong, often incrassate, in the basal row 2-4 inflated, to 80 μm high. Autoicous. Perigonial leaves about 0,3 mm long, ovate, abruptly acuminate. Perichaetial leaves 1.5-3 mm long, lanceolate with a long, flexuose, serrate acumen. Seta smooth, reddish, to 2.5 cm long; capsule inclined, ovoid, operculum conic, long-rostrate; peristome with the characters of the genus, exostome teeth about 400 μm long, not furrowed, endostome segments broad, cilia well-developed.

D i s t r i b u t i o n : Venezuela, Guyana, Suriname.

E c o l o g y : Epiphytic on tree trunks and branches or on decaying wood, occasionally on litter over rocks. Apparently rare and confined to higher altitudes. Not seen from French Guiana.

S e l e c t e d   s p e c i m e n s : Guyana: Upper Mazaruni Distr., trail from Kamarang to Pwipwi Mt., Alt. 650 m, Gradstein 5678 (U). Suriname: Brownsberg, Plateau road, Alt. 500 m, Florschütz 4676 (U); Bakhuis Mts., Alt. 550-700 m, Florschütz & Maas 2993 (U).

N o t e : The papillae in this species are very low and usually only visible in side-view on the concave part of the leaf back. If the papillae are over-looked the specimen could be taken for a species of *Sematophyllum*. It can be distinguished from *S. subsimplex* by the larger size of the elliptic leaves with long-acuminate apex and by the long perichaetial leaves with coarsely serrate acumen. From *S. galipense* it is separated by the long-acuminate, flexuose leaf apex.

2. **Trichosteleum hornschuchii** (Hampe) Jaeg., Ber. S. Gall. Naturw. Ges. 1876-77: 418. 1878. – *Hypnum hornschuchii* Hampe, Icon. Musc. 9 (annotation). 1844. – *Hypnum microcarpum* Hornsch., Fl. Bras. 1(2): 84 (Tab.4, fig.3). 1840, hom. illeg. Type: Brazil, Pará; Martius s.n., hb. Hooker 880 (BM).   – Fig. 165.

Trichosteleum fluviale (Mitt.) Jaeg., Ber. S. Gall. Naturw. Ges. 1876-77: 419. 1878, syn. nov. – *Sematophyllum fluviale* Mitt., J. Linn. Soc. Bot. 12: 493. 1869. Type: Colombia, Rio Magdalena, Weir 385 (NY).
Trichosteleum martii (C. Müll.) Kindb., Enum. Bryin. Exot.: 38. 1888, syn. nov. – *Hypnum martii* C. Müll., Linnaea 39: 466. 1875. – *Hypnum martianum* Lor., Moosstud.: 166. 1864, hom. illeg. Type: Suriname, Paramaribo, Wullschlägel 1243 (BR).
Trichosteleum subdemissum (Besch.) Jaeg., Ber. St. Gall. Naturw. Ges. 1876-77: 418. 1878, syn. nov. – *Rhaphidostegium subdemissum* Schimp. ex Besch., Ann. Sci. Nat. Bot. sér. 6, 3: 250. 1876. Type: Guadeloupe, L'Herminier s.n. (NY).

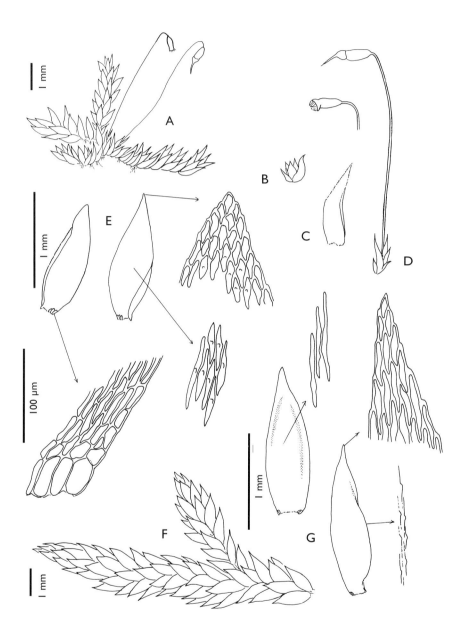

Fig. 165. *Trichosteleum hornschuchii* var. *hornschuchii*: A. stem portion with branches and capsules; B. perigonium; C. perichaetial leaf; D. perichaetium with capsule; E. branch leaves. *T. hornschuchii* var. *subglabrum*: F. branches; G. branch leaves. (A-E: Florschütz-de Waard & Zielman 5477; F-G: Cornelissen & ter Steege C 208).

Trichosteleum micropyxidium (– *"pyxis"*) (C. Müll.) Broth., Nat. Pfl. Fam.
1(3): 1119. 1908, syn. nov. – *Aptychus micropyxis* C. Müll., Malpighia 10:
518. 1896. Type: Guyana, Marshall Falls, Quelch s.n., hb. Levier 1272 (BM).

Dull green to bronze-green plants with light green innovations,
growing in flat to rough mats. Stems creeping or prostrate, irregularly
branched, branches usually short and prostrate with complanate
leaves, but in var. *subglabrum* more elongate and ascending with
imbricate leaves. Leaves ovate or oval-elliptic, 0.7-1.9 mm long and
0.2-0.7 mm wide, apex acute to rounded-acute, occasionally gradually
acuminate, margin crenulate or bluntly serrulate at apex; leaf cells thin
walled or incrassate, variable in size, at midleaf elongate-rhomboidal
to flexuose-linear, 40-100 μm long and 5-8 μm wide, at apex short,
irregularly rhomboidal or elliptic, papillae distinct or indistinct, sometimes very obscure and only visible in young leaves, basal cells
strongly incrassate and porose, alar cells rounded-quadrate to oval, in
the basal row conspicuously inflated and coloured, variable in size,
60-110 μm long and about 35 μm wide.
Autoicous. Perigonial leaves about 0.3 mm long, ovate, short-acuminate,
irregularly serrate. Perichaetial leaves 1-1.5 mm long, oblong with
acuminate, serrate apex, smooth. Seta smooth, reddish, 0.5-1.5 cm long;
capsule horizontal to pendent, ovoid, operculum long-rostrate; peristome
with the characters of the genus, exostome teeth 200-350 μm high, with
a broad median furrow and strongly papillose at apex, endostome segments
broad, finely papillose, cilia slender.

D i s t r i b u t i o n : West Indies, Guatemala, tropical South America; also
one record of tropical Africa.

N o t e s : This species was described from Brazil as *Hypnum microcarpum*
Hornsch. (hom. illeg.) and later renamed by Hampe as *Hypnum
hornschuchii*. Mitten (1869) erroneously mentioned it as a synonym of
Trichosteleum ambiguum (Schwaegr.) Par., but the type of that species,
Leskea ambigua (G) does not belong in *Trichosteleum*, having a single
costa extending half the leaf length. Most Spruce collections of *T.
hornschuchii* (NY, hb. Mitten) are identified as *T. ambiguum* and this
name has overruled the correct name. It is rehabilitated here.
T. fluviale (Mitt. 1968) is described as a small form of *T. hornschuchii*,
distinguished by smaller leaves and more distinct papillae. However, the
types are not essentially different and the differences all fit within the
extensive variation of this species.
In some collections the cells are more incrassate and narrower and the
papillae can be indistinct or totally absent. This makes it difficult to
recognize a *Trichosteleum* species in these collections; only specimens

with somewhat better perceptable papillae make the relationship clear. Also the larger size and the more imbricate, concave leaves give the plants a different aspect, which makes it useful to distinguish a variety.

2a. var. **hornschuchii**

Characteristic for the typical variety are the ovate leaves with thin walled, rhomboidal cells in the acute apex. The leaf length varies from 0.7-1.2 mm, but in a single collection leaf lengths to 1.7 mm occur. The papillae are low but distinct even in specimens growing on boulders in running water, the typical habitat for var. *subglabrum*. Both varieties can have leaves with rounded apex in that habitat.

E c o l o g y : Epiphytic on tree trunks and roots or on decaying wood in moist forest, occasionally on boulders in creeks. Common in Suriname, less frequent in Guyana and French Guiana.

S e l e c t e d   s p e c i m e n s : Guyana: Mabura Hill, 180 km SSE of Georgetown, Alt. 0-50 m, Cornelissen & ter Steege C198 (U); Kanuku Mts., Rupununi River-Puwib River, Alt. 80-100 m, Jansen-Jacobs et al. 225B (U). Suriname: Para River, swampy savanna near Hannover, Geijskes s.n. (3-12-1950) (U); Area of Kabalebo Dam project, Baruba creek, Alt. 1-50 m, Florschütz-de Waard & Zielman 5478 (U). French Guiana: Haut Tampoc River, Saut Awali, Cremers 4767 (CAY, U); Crique Pain de sucre, tributary of Comté River, Florschütz-de Waard 5962 (U).

2b. var. **subglabrum** J. Florsch., var. nov. Type: Guyana, Mabura Hill, Cornelissen & ter Steege 208 (U).

Ab var. *hornschuchii* differt: ramulis pluribus elongatis, foliis ovalis-ellipticis, cellulis linearibus, incrassatis et papillis inconspicuis vel absentibus.

The firm branches with imbricate, concave leaves give this variety a different aspect. The oval-elliptic leaves are 1-2 mm long and the leaf apex is often rounded, probably caused by the wet habitat. The linear, incrassate leaf cells with hardly visible papillae make it difficult to recognize this variety as *T. hornschuchii*. It is advised to look for papillae on young leaves in side view.

Collections of this variety frequently have been misidentified as *Sematophyllum subpinnatum*, but in that species the regularly incrassate leaf cells with fusiform lumina are determinant.

Ecology: On logs and boulders in marsh forest and creeks, also terrestrial on creek banks, temporarily submerged. Not common, restricted to a wet habitat.

Selected specimens: Guyana: Mabura Hill, dry evergreen forest on white sand, on clay and wood near creek, Cornelissen & ter Steege C208 (U, type); Upper Mazaruni Distr., trail from Kamarang to Pwipwi Mt., on boulders in river, Alt. 800 m, Gradstein 5736 (U). Suriname: Wana creek, trail from Moengo tapoe to Grote Zwiebelzwamp, Lanjouw & Lindeman 619 (U); Saramaca River, forest trail near Louisdam, Florschütz 1200 (U). French Guiana: Ile de Cayenne, Riv. Tonegrande, Cremers 6307 (CAY, U).

3. **Trichosteleum intricatum** (Thér.) J. Florsch., Trop. Bryol. 3: 98. 1990. – *Acroporium intricatum* Thér., Ann. Bryol. 7: 159. 1934. Type: French Guiana, St. Jean de Maroni, Gouv. Rey s.n., coll. Galliot (PC, NY). – Fig. 166.

Light green to bronze green plants, growing in loose mats. Stems elongate, creeping, sparingly to pinnately branched, branches procumbent or in horizontal rows extending from tree trunks, complanate-foliate. Leaves patent, lanceolate, to 1.5 mm long and 0.5 mm wide, apex slenderly acute or gradually acuminate, margin entire or slightly serrulate in upper part, usually narrowly revolute below apex at one or both sides; leaf cells oblong to linear, incrassate, at midleaf 30-70 µm long and 4-6 µm wide, shorter towards apex, shorter and wider towards base, strongly pitted and orange-coloured at insertion, alar cells small, rounded-quadrate or oval, incrassate, in the basal row elongate, the outer 2-5 inflated, 35-70 µm high and about 20 µm wide. Dioicous. Perigonia small with ovate leaves. Perichaetial leaves 1-1.7 mm long, slenderly lanceolate, serrulate, smooth. Seta smooth, reddish, 1-1.5 cm long; capsule inclined to pendent, ovoid with a wide mouth, operculum conic, obliquely long-rostrate; peristome with the characters of the genus, exostome teeth to 350 µm high, slightly transversely striolate in basal part, not furrowed, endostome segments broad, cilia well-developed, often double.

Distribution: Brazil, the Guianas.

Ecology: Epiphytic on branches and tree trunks in light forest. Not common.

Selected specimens: Guyana: Basin of Kuyuwini River, about 150 miles from the mouth, A.C. Smith 2520 (BM, U); Moraballi creek near Bartica, Richards 377A (BM, U). Suriname: Marowijne River, in forest

432

at base of Nassau Mts., Lanjouw & Lindeman 2241 (U); Area of Kabalebo Dam project, *Eperua* forest, Florschütz-de Waard & Zielman 5410 (U). French Guiana: Macouria, Michel, s.n. (NY); E-NE de Saül, 11 km N de Saut Maïs, Cremers 6259 (CAY, U).

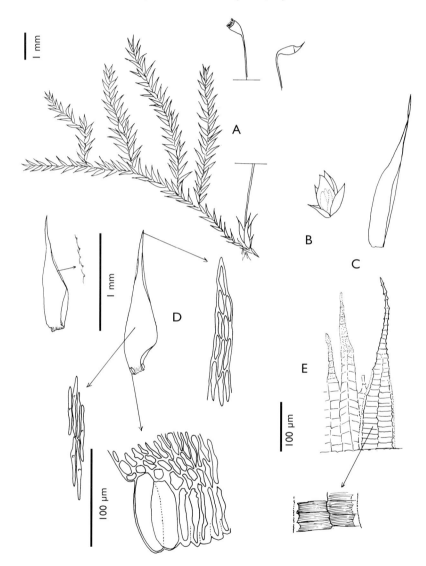

Fig. 166. *Trichosteleum intricatum*: A. stem with branches and capsules; B. perigonium; C. perichaetial leaf; D. branch leaves; E. part of peristome and detail of exostome tooth. (A, C-E: Florschütz 433; B: Florschütz-de Waard & Zielman 5394).

Notes: Thériot described this species as *Acroporium*; the patent-spreading leaves are reminiscent of *Acroporium pungens*, but the alar cells are small and the presence of papillae, although small and indistinct, prove the relationship with *Trichosteleum*.

The species is best recognized by the habit with elongated stems and many branches of the same length, with patent-spreading, lanceolate leaves. As the papillae are easily overlooked the species is often misidentified as *Sematophyllum subsimplex*. It is different in the more incrassate leaf cells and the little-inflated, often incrassate alar cells. It is only reported from Brazil and the Guianas, but it might prove to be more wide-spread if better recognized.

4. **Trichosteleum papillosum** (Hornsch.) Jaeg., Ber. S. Gall. Naturw. Ges. 1876-77: 419. 1878. – *Hypnum papillosum* Hornsch., Fl. Bras. 1(2): 82. 1840. Type: Brazil, Minas Geraïs, Beyrich s.n., Hb. Hooker (BM, NY). (The collector is not indicated on the label, but the detailed collecting data correspond with those in the description).

– Fig. 167.

Hypnum spirale C. Müll., Linnaea 21: 200. 1848. Type: Suriname, Weigelt s.n. (BM).
Trichosteleum guianae (C. Müll.) Broth., Nat. Pfl., Fam. 1(3): 1119. 1908, syn. nov. – *Sigmatella guianae* C. Müll., Malpighia 10: 518. 1896. Type: Guyana, Marshall Falls, Quelch s.n., hb. Levier 1271 (BM).

Light green to yellowish green plants growing in flat or rough mats. Stems creeping, defoliate with age, sparingly divided to irregularly pinnate, branches prostrate, 1/2-2 cm long. Leaves slightly complanate, sometimes homomallous, elliptic to lanceolate, 0.7-1.8 mm long and 0.2-0.5 mm wide, apex often abruptly acuminate, acumen slender or broad, usually flexuose, crispate when dry, margin serrate in upper part, serrulate towards base, often narrowly reflexed below apex; leaf cells incrassate, with a coarse, blunt papilla on both surfaces, except in the basal and apical cells, papillae to 10 μm high, with a broad base, as wide as the cell lumen, midleaf cells elongate-rhomboidal to linear, 40-85 μm long and 5-7 μm wide, shorter and strongly incrassate towards base, alar cells rounded-quadrate or rectangular, in the basal row 2 to 4 strongly inflated, 60-100 μm long and about 30 μm wide.

Synoicous. Perichaetial leaves smooth, slenderly lanceolate, 1-1.5 mm long, long-acuminate, serrate-dentate. Seta reddish, smooth throughout or rough in upper part, 0.5-2 cm long; capsule horizontal to pendent, with a distinct rough neck with stomata, operculum long-rostrate; peristome with the characters of the genus, exostome teeth to 350 μm long, furrowed, strongly papillose at apex, endostome segments very broad, cilia single or double.

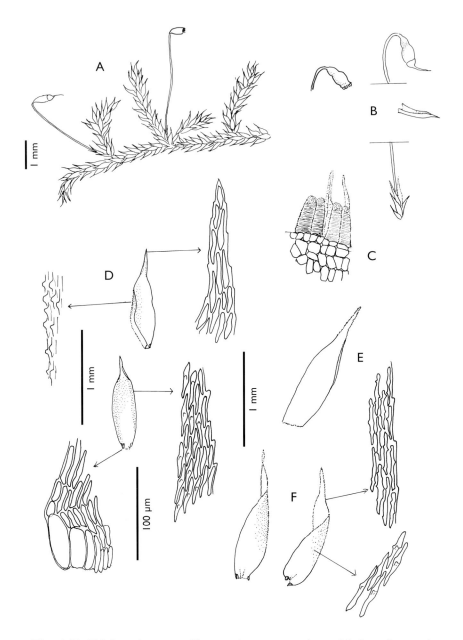

Fig. 167. *Trichosteleum papillosum*: A. stem portion with branches and capsules; B. perichaetium, capsules and calyptra; C. part of peristome with incurved exostome teeth; D. branch leaves; E. perichaetial leaf; F. branch leaves. (A-D: Florschütz 4801; E-F: Lanjouw & Lindeman 2899).

Distribution: Brazil, Colombia, the Guianas.

Ecology: Epiphytic on bark of living trees and on decaying wood. A common species in lowland rain forest and savanna forest, not collected in the coastal area.

Selected specimens: Guyana: Mabura Hill near Yaya creek, Alt. 0-50 m, Cornelissen & ter Steege C095 (U); N slope of Mt Roraima, Alt. 1200-1600 m, Gradstein 5316 (U). Suriname: Distr. Nickerie, Area of Kabalebo Dam project, Alt. 0-50 m, Florschütz-de Waard & Zielman 5312 (U); Brownsberg, Mazaroni plateau, Alt. 500 m, Florschütz 4801 (U). French Guiana: Mt. de l'Inini, Alt. 700 m, Cremers 9220 (CAY, U); Eaux Claires, 5 km N of Saül, Alt. 100-200 m, Florschütz-de Waard 5896 (U).

Notes: This species is easily recognized by the conspicuous papillae and the flexuose, acuminate leaf tip, crispate when dry. The serrate upper leaf margin is a rather constant character. Hornschuch described the margin as "obscure serrulata aut integerrima" but the type has distinctly serrate margins in the leaf apices.
A variable character is the shape of the leaf tip, which can be slenderly acuminate with strongly reflexed margins or more gradually acuminate and flat; this can vary along the same plant. The differences in size of this species are surprising: very slender specimens occur, growing in flat mats, with short branches and small leaves to 1 mm long, but also robust specimens with more elongated branches and leaves of at least 1.5 mm long, forming dense tufts. The latter specimens have more strongly incrassate cell walls and lower papillae. No discontinuous patterns of variation could be observed to determine varieties. Ecological factors could be responsible for the variation: the small plants are usually collected from decaying logs in moist rainforest, whereas the larger plants were growing on logs in savanna forest or epiphytic on branches in the canopy.
T. sentosum (Sull.) Jaeg., a species known from Central America and the West Indies, is very close to *T. papillosum*; the type specimen of Cuba (FH) differs in the very coarse papillae (to 20 μm high) and the long and slender, sharply dentate leaf apex. Considering the variation in these characters it could be an extreme form of *T. papillosum*.

8. **WIJKIA** Crum, Bryologist 74: 170. 1971.
Lectotype (Crum 1971): W. extenuata (Brid.) Crum (Hypnum extenuatum Brid).

Acanthocladium Mitt. p.p., Trans. Roy. Soc. Victoria 19: 85. 1883. – *Acanthocladium* sensu Broth., in Engl. & Prantl, Nat. Pfl. Fam. (ed.2) 11: 412. 1925.

Pale green plants growing in dense mats. Stems creeping or prostrate, branches often long and ascending, sometimes flagellate, irregularly to bipinnately divided; pseudoparaphyllia present. Stem and branch leaves differentiated, stem leaves ovate, branch leaves smaller, ovate-lanceolate; costa short and double, indistinct; leaf cells elongate-hexagonal to linear, usually smooth.
Seta long and twisted, smooth; capsule inclined, exothecial cells not collenchymatous, operculum rostrate; peristome double, exostome teeth transversely striolate in basal part, papillose in upper part, endostome segments from a high basal membrane, keeled, cilia present.

N o t e : For taxonomic comments on this genus see Crum (1971) and Buck (1986).

D i s t r i b u t i o n : pantropical, only one species in the Guianas.

1. **Wijkia costaricensis** (Dixon & Bartr.) Crum, Bryologist 74: 170. 1971. – *Acanthocladium costaricense* Dixon & Bartr., J. Washington Ac. Sc. 21: 294. 1931. Type: Costa Rica, Prov. San José, Standley & Valerio 43395 (FH, NY).   – Fig. 168.

Yellowish green plants growing densely interwoven in rough mats. Stems creeping, freely branched, branches ascending, often curved, irregularly pinnate or bipinnate, proliferous; pseudoparaphyllia foliose. Stem leaves imbricate, asymmetric to subfalcate, often homomallous, ovate, to 1.2 mm long and 0.5 mm wide, apex acute or short-acuminate, margin often partly reflexed, distantly serrulate at apex; branch leaves erect-spreading, often complanate, ovate-lanceolate with acute apex and irregularly serrate or dentate margin, 0.3-0.8 mm long, at base of the branches often much smaller and with rounded apex; leaf cells smooth, slightly incrassate, at midleaf flexuose-linear, 40-60 μm long (to 80 μm in stem leaves) and 5-6 μm wide, in apex shorter, elongate-rhomboidal, at leaf base shorter and wider with porose walls, alar cells rounded-quadrate or rectangular in a more or less triangular group, in the basal row coloured and sometimes conspicuously inflated, 35-55 μm long and about 30 μm wide, often transversely divided.

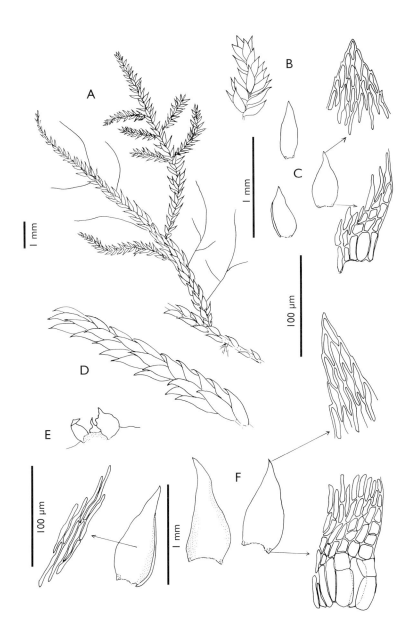

Fig. 168. *Wijkia costaricensis*: A. stem portion with bipinnate branch; B. end of branchlet; C. branch leaves; D. stem portion; E. pseudoparaphyllia; F. stem leaves. (A: Florschütz 4848; B-F: Cremers 9243).

Sporophyte not seen and apparently undescribed for the species.

Distribution: Central America, Colombia, the Guianas.

Ecology: Epiphytic on tree trunks and branches. In the Guianas rare and confined to higher altitudes, not seen from Guyana.

Specimens examined: Suriname: van Asch van Wijck Mts., summit Ebbatop, Alt. 700 m, Florschütz 1504 (U); Lely Mts., plateau 5, Alt. 650 m, Florschütz 4848 (U).

Note: A moss with a characteristic growth form, resulting from the ascending, often curved branches, irregularly pinnate or bipinnate and at the end often continuing growth as a stem. The older stems are often more or less "hypnoid" by the homomallous stem leaves. Stem and branch leaves are differentiated in size and position, but not strictly dimorphous as described by Dixon and Bartram. The branch leaves are erect-spreading and narrower than the imbricate stem leaves but the shape is not essentially different; along the primary and secondary branches all intermediates occur.

HYPNACEAE

by

J. Florschütz-de Waard

and

K. Veling[7]

Small to robust plants. Stems creeping, irregularly to pinnately branched, usually with central strand; pseudoparaphyllia present. Leaves complanate or homomallous, often asymmetric, ovate to lanceolate, usually acuminate, costa short and double or lacking; leaf cells mostly linear, sometimes rhomboidal or hexagonal, smooth or prorate or papillose, alar cells usually differentiated, often quadrate, seldom inflated.

Autoicous or dioicous. Seta elongate; capsule inclined to pendent, exothecial cells usually not collenchymatous, operculum conic to short-rostrate; peristome double, exostome teeth not furrowed, lanceolate, transversely striolate below, papillose at apex, endostome segments from a high basal membrane with well-developed cilia. Calyptra cucullate.

N o t e : The delimitation of this family has gone through many changes during the last decades. Repeatedly genera have been added to or excluded from the Hypnaceae. *Isopterygium*, previously placed in the Plagiotheciaceae, is now back in the Hypnaceae (Ireland 1992). *Phyllodon* (*Glossadelphus*), excluded from the Sematophyllaceae by Seki (1969), is now definitely transferred to the Hypnaceae by Nishimura et al. (1984). The latter authors delimited the Hypnaceae from other pleurocarpous families and emphasized the following characters as distinguishing with regard to the Sematophyllaceae: leaf shape usually asymmetric, alar cells not vesiculate-inflated, exothecial cells not collenchymatous, operculum conic-obtuse to short-rostrate (in Sematophyllaceae long-rostrate or subulate).

Buck (1984) divided *Mittenothamnium* into 2 genera, leaving the species with ascending secondary stems in *Mittenothamnium* and transferring the species with prostrate stems to *Chrysohypnum*, an older name for the genus.

[7] Herbarium Division, Department of Plant Ecology and Evolutionary Biology. Heidelberglaan 2, 3584 CS Utrecht, The Netherlands.

LITERATURE

Bartram, E.B. 1949. Mosses of Guatemala. Hypnaceae. Fieldiana Bot. 25: 397-415.

Brotherus, V.F. 1901. Report on two botanical collections made by Messrs. F.V. McConnell and J.J. Quelch at Mount Roraima in British Guiana. Trans Linn. Soc. London, Bot. ser. 2.6; Musci: 88-93.

Buck, W.R. 1983. Nomenclatural and taxonomic notes on West Indian Sematophyllaceae. Brittonia 35: 309-311.

Buck, W.R. 1984. Taxonomic and nomenclatural notes on West Indian Hypnaceae. Brittonia 36: 178-183.

Buck, W.R. 1987. Notes on Asian Hypnaceae and associated taxa. Mem. New York Bot. Gard. 45: 519-527.

Buck, W.R. 1993. Taxonomic results of the BRYOTROP expedition to Zaïre and Rwanda 24. Trop. Bryol. 8: 199-217.

Crum, H.A. 1994. Rhacopilopsis in A.J. Sharp, H.A. Crum & P.M. Eckel. The Moss Flora of Mexico 2. Mem. New York Bot. Gard. 69: 1049-1051.

Dixon, H.N. 1922. Rhacopilopsis trinitensis Britt. & Dixon. J. of Bot. 60: 86-88.

Ireland, R.R. 1992. The moss genus Isopterygium in Latin America. Trop. Bryol. 6: 111-132.

Iwatsuki, Z. & M.R. Crosby. 1979. Lectotypification of the genus Isopterygium Mitt. J. Hattori Bot. Lab. 45: 389-393.

Mitten, W. 1869. Musci Austro-Americani. J. Linn. Soc. Bot. 12, Stereodonteae: 497-537.

Müller, C. 1851. Synopsis Muscorum Frondosorum II. Berlin.

Nishimura, N. & M. Niguchi, T. Seki, H. Ando. 1984. Delimitation and subdivision of the moss family Hypnaceae. J. Hattori Bot. Lab. 55: 227-234.

Nishimura, N. & H. Ando. 1994. Ectropothecium in: A.J. Sharp, H.A. Crum & P.M. Eckel. The Moss Flora of Mexico 2. Mem. New York Bot. Gard. 69: 1037-1039.

Robinson, H. 1965. Notes on Oreoweisia and Hypnella from Latin America. The Bryologist 68: 331-334.

Seki, T. 1968. A revision of the family Sematophyllaceae of Japan with special reference to a statistical demarcation of the family. J. Sci. Hiroshima Univ., ser. b, div. 2, 12: 1-80.

Steyermark, J.A. 1981. Erroneous citations of Venezuelan localities. Taxon 30: 816-817.

KEY TO THE GENERA

1 Midleaf cells oval to elongate-hexagonal, lax, 10-40 µm wide · · · · · · · · · · 2
  Midleaf cells linear, narrow, less than 10 µm wide · · · · · · · · · · · · · · · · · · · 3

2 Midleaf cells 30-100 µm long, 10-18 µm wide. Capsule with conic, short-rostrate or apiculate operculum, exostome teeth not furrowed · · · · · · · · · ·
  · *7. Vesicularia*
  Midleaf cells 100-200 µm long, 10-40 µm wide. Capsule with slenderly rostrate operculum, exostome teeth with a median furrow · · · · · · · · · · · · ·
  · (see *Leucomiaceae*, p)

3 Leaf cells smooth (sometimes with a luminous spot at cell ends, but in side view not papillose) · 4
  Leaf cells with papillae or projecting cell ends (in side view visible as distinct papillae) · 6

4 Branch leaves dimorphous, arranged in 4 rows: dorsal and lateral leaves complanate-spreading in 2 rows, asymmetric to falcate; ventral leaves erect in 2 rows, subsymmetric · *6. Rhacopilopsis*
  Branch leaves different, not arranged in 4 rows · 5

5 Plants subpinnately branched; leaves usually strongly falcate and homomallous, serrate at apex. Perichaetia conspicuous, to 3.5 mm high · · · · · · · · · · · · ·
  · *2. Ectropothecium*
  Plants irregularly branched; leaves symmetric to slightly falcate, complanate or erect-spreading, bluntly serrulate at apex. Perichaetia small, to 1.5 mm high · *3. Isopterygium*

6 Leaf apex blunt or truncate, leaf cells with 1-4 papillae · · · · · · *5. Phyllodon*
  Leaf apex acute or acuminate, leaf cells with projecting cell ends · · · · · · · 7

7 Plants with ascending, curved secondary stems with stipitate base. Leaf cells projecting as papillae at distal end only · · · · · · · · · · · *4. Mittenothamnium*
  Plants with prostrate secondary stems, not stipitate. Leaf cells projecting as papillae at both ends · *1. Chrysohypnum*

1. **CHRYSOHYPNUM** Hampe, Bot. Zeit. 28: 35. 1870.
Type: C. patens Hampe

Small to medium-sized plants in often dense mats. Stems creeping, irregularly to pinnately branched; pseudoparaphyllia filamentous to narrowly lanceolate. Stem and branch leaves slightly differentiated, sub-symmetric, stem leaves ovate to ovate-triangular, acuminate, margin serrulate, costae double, to 1/3 of leaf length, branch leaves narrower, ovate to ovate-lanceolate, margin more distinctly serrulate; leaf cells linear, projecting as papillae at both ends, alar cells more or less differentiated. Autoicous. Seta elongate, smooth; capsule ovoid-cylindric, inclined to pendent, operculum conic-obtuse or conic-rostellate; peristome double, exostome teeth densely striolate at base, minutely papillose at apex, endostome consisting of a high basal membrane with keeled, papillose segments and cilia.

Distribution: pantropical, only one species in the Guianas.

1. **Chrysohypnum diminutivum** (Hampe) Buck, Brittonia 36: 182. 1984. – *Hypnum diminutivum* Hampe, Linnaea 20: 86. 1847. – *Mittenothamnium diminutivum* (Hampe) Britt., Bryologist 17: 9. 1914. Type: Venezuela, Caracas, Moritz 20 (BM).   – Fig. 169.

Small, green to yellowish green plants with creeping stems, subpinnately branched, branches prostrate or ascending; pseudoparaphyllia filamentous to slenderly lanceolate. Stem leaves patent-spreading, ovate to broad-ovate, gradually to abruptly acuminate, 0.5-1 mm long and 0.25-0.45 mm wide, smaller on young stems, margin serrulate, costae extending 1/4-1/3 of leaf length, sometimes faint; branch leaves patent-spreading or complanate, ovate to ovate-lanceolate, acute to acuminate, 0.65-1 mm long and 0.25-0.4 mm wide, margin serrate nearly to base; leaf cells elongate-rhomboidal to linear, projecting at both cell ends on dorsal side of leaf, 25-60 µm long and 3-5 µm wide, alar cells quadrate to rectangular, forming a small but distinct group.

Autoicous. Perichaetia and perigonia produced on stem; perigonial leaves ovate to orbicular, short-acuminate; inner perichaetial leaves triangular, to 1 mm long, slenderly acuminate. Seta reddish, 1.5-2 cm long, smooth; capsule inclined, oval, 1 mm long, operculum conic, mamillate; peristome with the characters of the genus, exostome teeth 300-400 µm long, endostome segments of the same length, cilia 1-2, very fragile. Calyptra narrowly cylindric.

Distribution: Southern U.S., West Indies, Central and tropical South America.

Ecology: Epiphytic on bark of trees, occasionally on logs, also epilithic or terrestrial. Growing preferably in light habitats: cultivated areas, low savanna forest, sometimes in lowland rainforest, also on higher altitudes.

Selected specimens: Guyana: Kanuku Mts., A.C. Smith 3197 (BM, U); Upper Mazaruni district, Jawalla, Alt. 500 m, Gradstein 4944

(U). Suriname: Kabalebo Dam area, shrub savanna, Florschütz-de Waard & Zielman 5480. French Guiana: 2 km SW of Saül, Alt. 220 m, Montfoort 473 (U); Saül, Cremers 4213 (CAY, U).

N o t e : *Chrysohypnum diminutivum* is best recognized by the leaf cells with papillae at both cell ends. It can be easily distinguished from *Mittenothamnium reptans* by the ovate stem leaves; for more detailed differences see under that species.

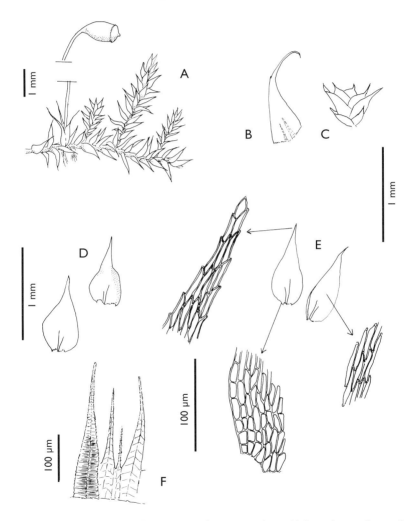

Fig. 169. *Chrysohypnum diminutivum*: A. stem portion with branches and capsule; B. inner perichaetial leaf; C. perigonium; D. stem leaves; E. branch leaves; F. exostome tooth with part of endostome (Florschütz-de Waard & Zielman 5480).

2. **ECTROPOTHECIUM** Mitt., J. Linn Soc. Bot. 10: 180. 1868.
  Type : E. tutuilum (Sull.) Mitt. (Hypnum tutuilum Sull.)

Small to medium-sized plants. Stems creeping, pinnately branched, branches prostrate; pseudoparaphyllia filiform or slenderly foliose. Stem and branch leaves more or less differentiated, often falcate-secund, acuminate, costae short and double or absent; leaf cells linear, smooth or faintly projecting at distal cell end, alar cells little differentiated.
Autoicous or dioicous. Seta elongate, smooth; capsule horizontal to pendent; peristome double, exostome teeth striolate in basal part, papillose in upper part, endostome with broad, keeled segments and cilia on a high basal membrane.

Distribution: pantropical; only one species in the Guianas.

1. **Ectropothecium leptochaeton** (Schwaegr.) Buck, Brittonia 35: 311. 1983. – *Hypnum leptochaeton* Schwaegr., Spec. Musc. Suppl. 1(2): 296. 1816. Type: Richard s.n., Cayenne, Richard s.n. (PC). – Fig. 170.

> *Ectropothecium globitheca* (C. Müll.) Mitt., J. Linn. Soc. Bot. 12: 512. 1869.
> – *Hypnum globitheca* C. Müll., Syn. 2: 300. 1851. Type: Venezuela, prope Galipan, Funk & Schlim s.n., coll. Linden nr. 350 p.p. (BM).
> *Ectropothecium apiculatum* Mitt., J. Linn. Soc. Bot. 12: 512. 1869. Type: Brasil, Minas Geraïs, Gardner 1196 (NY).
> *Ectropothecium guianae* Broth. & Par., Rev. Bryol. 33: 57. 1906, syn. nov. Type: French Guiana, Camp de la forestière, Maroni, Michel s.n. (PC).

Medium-sized, bright-green, shiny plants, growing in loose or dense mats. Stems creeping, regularly or irregularly pinnate-branched; pseudoparaphyllia filamentous to slenderly lanceolate. Stem- and branch leaves often strongly falcate-secund, complanate-spreading or homomallous, stem leaves ovate-lanceolate to lanceolate, short- or long-acuminate, 1-1.6 mm long and 0.2-0.5 mm wide, branch-leaves smaller, ovate to lanceolate, 0.8-1.3 mm long and 0.2-0.4 mm wide, costae extending to 1/3 of leaf length, often faint, margin plane, serrulate in upper half, occasionally nearly entire; leaf cells linear, to 80 µm long and 5 µm wide, laxer towards base, alar cells slightly differentiated with 1 or 2 enlarged cells in the basal row, often remaining attached to the stem.
Autoicous. Perigonial leaves ovate, to 0.5 mm long, short-acuminate. Perichaetial leaves lanceolate, to 3.5 mm long, gradually long-acuminate, subentire to serrulate in upper part. Seta smooth, 1.2-1.5 cm long; capsule pendent, ovoid, 1 mm long, constricted below the mouth when dry, operculum short-rostrate; peristome with the characters of the genus, exostome teeth 400 µm long, endostome segments minutely papillose, cilia 1-2. Calyptra narrowly cylindric.

Distribution: West Indies, Central and tropical South America.

Ecology: In moist rainforest, on bark of trees and on decaying wood, occasionally on stones. Rather common in Suriname and French Guiana, seldom collected in Guyana.

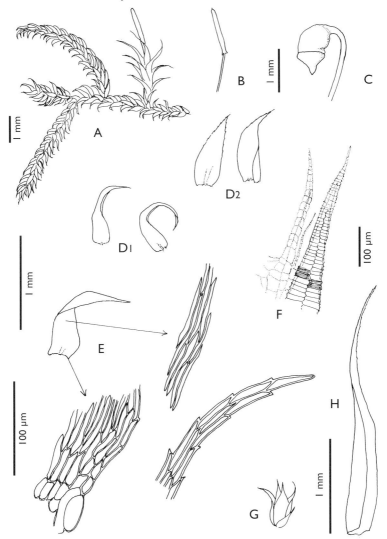

Fig. 170. *Ectropothecium leptochaeton*: A. stem portion with branches and perichaetium; B. immature capsule with calyptra; C. full-grown capsule; D1+D2. branch leaves; E. stem leaf; F. exostome tooth with part of endostome; G. perigonium; H. inner perichaetial leaf. (A-D1, E, G-H: Florschütz-de Waard & Zielman 5347; D2: ibid 5059; F: Arnoldo 3506).

Selected specimens: Guyana: Mazaruni distr., Jawalla, Alt. 500 m, Gradstein 4932 (U). Suriname: Upper Saramaca River, near Louisdam, Florschütz 1205 (U); Kabalebo Dam project, W bank of Kabalebo River, Florschütz-de Waard & Zielman 5347 (U). French Guiana: Mt. de l'Inini, Alt. 670-800 m, Cremers 9068 (CAY, U); Eaux Claires, 5 km N of Saül, Alt. 100-200 m, Florschütz-de Waard 5887 (U).

Notes: The variation in leaf shape in this species is confusing, varying from ovate and short-acuminate to lanceolate and long-acuminate. Most collections, however, are easily recognized by the usually strongly falcate leaves, homomallous at the end of the branches. The conspicuous high perichaetia also form a good distinguishing character.

Buck (1983) reported that *Hypnum leptochaeton*, described from Cayenne and long considered to belong to the Sematophyllaceae, actually is an *Ectropothecium*. The name *E. leptochaeton* has priority over all other names that in the course of time have been used for this species; this made *E. apiculatum* Mitt. synonymous. Bartram (1949) had already concluded that *E. apiculatum* and *E. globitheca* are "uncomfortably close". Examination of the types of both species made it clear that the differences in leaf shape are all included in the continuous variation in the numerous collections from the Guianas. This corresponds with the opinion of Nishimura & Ando (1994). In addition *E. guianae* described from French Guiana proved to belong to *E. leptochaeton*.

3. **ISOPTERYGIUM** Mitt., J. Linn. Soc. Bot. 12: 21. 1869.
   Lectotype (Iwatsuki & Crosby 1979): I. tenerum (Sw.) Mitt. (Hypnum tenerum Sw.).

Small to medium-sized plants in thin to dense mats. Stems creeping, sparingly and irregularly branched; pseudoparaphyllia filiform. Stem and branch leaves not differentiated, erect-spreading to complanate, symmetric or asymmetric, ovate to lanceolate, apex acute to acuminate, margin usually serrulate in upper half, costa lacking or short and double; leaf cells smooth, linear, alar cells more or less differentiated, quadrate to rectangular.

Autoicous or dioicous. Seta elongate, smooth; capsule erect, inclined or pendent, operculum conic to short-rostrate, exothecial cells often slightly collenchymatous; peristome double, exostome teeth striolate at base, papillose at apex, endostome segments and cilia on a low to high basal membrane.

Distribution: pantropical and subtropical.

KEY TO THE SPECIES

1 Most leaves symmetric, slenderly lanceolate, 3-4 times as long as broad; apex acute to gradually acuminate. Mature capsules erect or inclined ········
·· *1. I. subbrevisetum*
Most leaves asymmetric to slightly falcate, ovate-lanceolate, 2-3 times as long as broad; apex acuminate. Mature capsules horizontal to pendent ···
·· *2. I. tenerum*

1. **Isopterygium subbrevisetum** (Hampe) Broth., Nat. Pfl. Fam. 1(3): 1081. 1908. – *Hypnum subbrevisetum* Hampe, Vidensk. Meddel. Naturhist. Foren. Kjoebenhavn. ser. 3,6: 165. 1875. Type: Brazil, vicinity of Rio de Janeiro, Glaziou 6356 (BM).   – Fig. 171.

Slender, light green plants often growing in dense mats. Stems creeping or prostrate, to 2 cm long, freely branched, branches ascending with erect-spreading leaves or prostrate with complanate-spreading leaves. Leaves lanceolate, mostly symmetric, long-acute or very gradually acuminate, 0.4-1.2 mm long and 0.1-0.3 mm wide, margin minutely serrulate or crenulate throughout, costa short and double, often indistinct; leaf cells linear, 70-110 µm long and 3-7 µm wide, shorter in leaf apex and towards base, alar cells little differentiated, usually consisting of a few subquadrate cells.
Autoicous. Perigonial leaves broadly ovate, acute or short-acuminate, leaf cells short. Perichaetial leaves lanceolate, to 1 cm long, acute, little differentiated. Seta smooth, reddish, 4-9 mm long; capsule erect to inclined, ovoid-cylindric, operculum conic, slenderly rostrate; peristome with the characters of the genus, exostome teeth to 200 µm high, endostome segments broad, fragile, minutely papillose, cilia 1 or 2, papillose.

Distribution: West Indies, Central and tropical South America.

Ecology: Epiphytic on bark of trees or rotten logs. Usually in open areas in moist forest. Not common in Suriname and French Guiana, not collected in Guyana.

Selected specimens: Suriname: Brownsberg, Alt. 450 m, Gradstein 4670 (U). French Guiana: Sommet sud du Pic Matecho NE of Saül, Alt. 590 m, Cremers 6291 (CAY, U); Eaux Claires, 5 km N of Saül, Alt. 100-200 m, Florschütz-de Waard 5953 (U).

Note: The differences between this species and *I. tenerum* are ill-defined. In the latter species branches with distant, slenderly lanceolate leaves could also be observed. The most reliable difference is in the habit (of well-developed specimens) with many ascending branches and narrow, stiff-spreading, symmetric leaves. The capsules, which are erect or in maturity slightly inclined, may also help to identify this species.

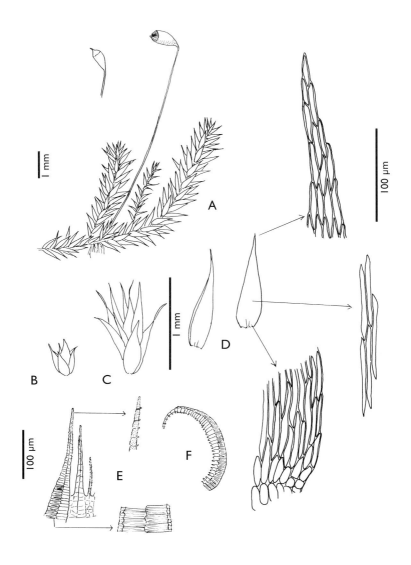

Fig. 171. *Isopterygium subbrevisetum*: A. stem portion with branches and capsules; B. perigonium; C. perichaetium; D. leaves; E. part of peristome with detail of exostome; F. exostome tooth, side view. (Florschütz-de Waard 5953).

2. **Isopterygium tenerum** (Sw.) Mitt., J. Linn. Soc. Bot. 12: 499. 1869.
– *Hypnum tenerum* Sw., Fl. Ind. Occ. 3: 1817. 1806. Type: Jamaica,
Swartz 2719 (BM). – Fig. 172.

Isopterygium radicisetum (C. Müll.) Broth., Nat. Pfl. Fam. 1(3): 1082. 1908.
– *Plagiothecium radicisetum* C. Müll., Malpighia 10: 515. 1896. Type:
Guyana, Marshall Falls, Quelch 1265 (NY, PC).
Ectropothecium isopterygioides Card. & Thér., Ann. Bryol. 7: 160. 1934, syn.
nov. Type: French Guiana, Maroni, Crique Balatée, Gouv. Rey (PC).

Slender, yellowish green plants, growing in thin mats. Stems creeping, to
3 cm long, irregularly branched; pseudoparaphyllia filiform; clustered
propagulae sometimes present on stem, filiform, simple or branched,
papillose. Leaves erect to complanate-spreading, stem leaves sometimes
homomallous, asymmetric to slightly falcate, ovate to ovate-lanceolate,
0.4-1.5 mm long and 0.2-0.5 mm wide, apex acuminate, margin usually
bluntly serrulate in upper half, occasionally entire throughout, costae
short, often unequal or absent; leaf cells linear, often flexuose, 55-140
µm long and 4-9 µm wide, apical cells shorter, basal cells shorter and
wider, oval-rectangular, often slightly inflated in the basal row, alar cells
little differentiated, short-rectangular or quadrate in a small group, sometimes lacking.
Autoicous. Perigonial leaves ovate, short-acuminate, to 0.25 mm long.
Perichaetial leaves lanceolate, acuminate, serrulate near apex, to 1.5
mm long. Seta yellowish to reddish, 5-15 mm long, smooth; capsule
pendent, ovoid-cylindric, arcuate, to 1 mm long, operculum conic,
short-rostrate; peristome with the characters of the genus, exostome
teeth about 200 µm long, endostome segments minutely papillose, cilia
1-2, papillose.

D i s t r i b u t i o n : Southern U.S., West Indies, Central and tropical
South America.

E c o l o g y : Epiphytic on decaying wood or bark of living trees,
occasionally epiphyllous; also terrestrial in moist places. In all kinds of
vegetation, preferrably in well illuminated, moist habitats, e.g. wet areas
in savanna vegetation and near creeks. Frequent in Suriname and locally
in Guyana, rare in French Guiana.

S e l e c t e d s p e c i m e n s : Guyana: Upper Mazaruni distr., Kamarang,
Pakaraima Mts., Alt. 500 m, Gradstein 4782 (U); Upper Mazaruni distr.,
savannah bush 5 km N of Maramadam, Alt. 650 m, Gradstein 5664 (U).
Suriname: Tibiti savanna, Lanjouw & Lindeman 1750 A (U);
Paramaribo, Cultuurtuin, Florschütz 4593 (U). French Guiana: 2 km SW
of Saül, Alt. 180-210 m, Montfoort & Ek 514 (U).

450

Note: This species is extremely variable in size and habit. In the collections from the Guianas very small specimens could be observed with leaves not over 0.6 mm long and on the other hand very large specimens with a leaf length to 1.5 mm. *Isopterygium tenerifolium*, reported by Ireland (1992) for Suriname and Guyana, is characterized by the large size of the plants with a leaf length of 1-1.5 mm and a seta

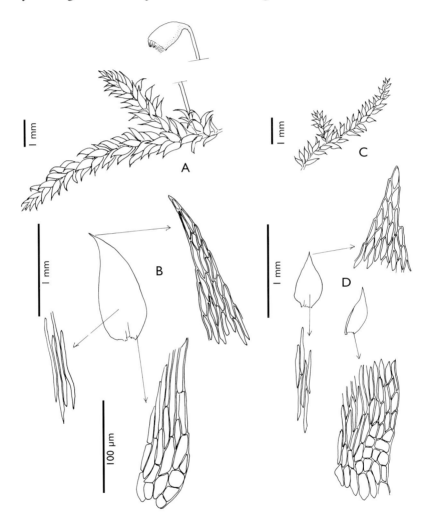

Fig. 172. *Isopterygium tenerum*: A. stem portion with branch and capsule; B. leaf; C. stem portion and branch; D. leaves. (A-B: Gradstein 5568; C-D: Gradstein 4782).

length of 2-3 cm. The collection mentioned as a possible representative of *I. tenerifolium* (Teunissen & Wildschut 11931) does not have sporophytes and in the variation of size no discontinuity could be observed. In all collections from the Guianas the setae are shorter than 2 cm therefore the occurrence of *I. tenerifolium* could not be confirmed.

4. **MITTENOTHAMNIUM** Henn., Hedwigia 41 (Beibl.): 225. 1902 (nom. cons.).
Type: M. reptans (Hedw.) Card. (Hypnum reptans Hedw.).

Microthamnium Mitt., J. Linn. Soc. Bot. 12: 503. 1869 (hom. illeg.).

Small to robust, often shiny plants. Primary stems creeping, secondary stems arched, with stipitate base, irregularly to bi- or tripinnately branched; pseudoparaphyllia foliose. Stem and branch leaves differentiated: stem leaves triangular to orbicular, margins serrulate, branch leaves ovate-lanceolate, margins serrate, costae short and double; leaf cells linear, papillose at back by projecting cell ends at distal corner.
Autoicous or dioicous. Seta elongate, smooth; capsule ovoid-cylindric, inclined to horizontal, operculum conic-rostrate; peristome double, exostome teeth densely striolate below, papillose above, endostome segments and cilia on a high basal membrane.

Distribution: Neotropics and tropical Africa; only one species in the Guianas.

1. **Mittenothamnium reptans** (Hedw.) Card., Rev. Bryol. 40: 21. 1913.
– *Hypnum reptans* Hedw., Spec. Musc.: 265. 1801. Type: Jamaica, et Insulae meridionales, Swartz s.n. (G). – Fig. 173.

Medium-sized, light green plants, growing in loose, sometimes extensive mats. Primary stems creeping, elongate, secondary stems ascending and arched, often rooting at the end, irregularly or pinnately to bipinnately branched, distantly foliate; pseudoparaphyllia triangular. Stem leaves patent, on the stipitate base of the secondary stems sometimes squarrose, triangular, acute-acuminate, 0.8-1.3 mm long and 0.3-0.5 mm wide, smaller towards base and end of stem, margin serrulate in upper half, costae unequal, extending 1/4 to 1/3 of leaf length, sometimes faint; branch leaves erecto-patent, often complanate-spreading, to 1.4 mm long and 0.5 mm wide, narrowly triangular to ovate-lanceolate, acute-acuminate, serrate in upper half; leaf cells linear, projecting at back on

452

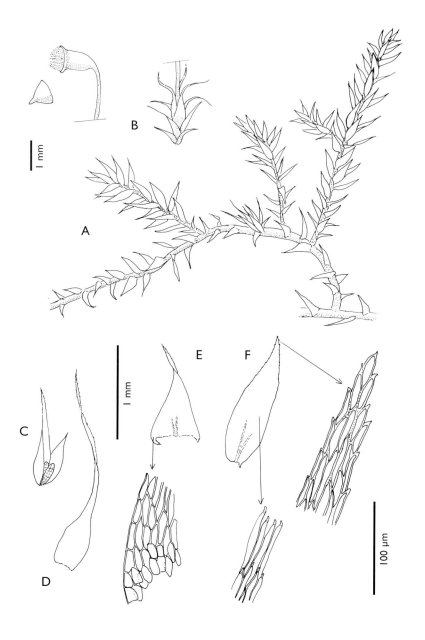

Fig. 173. *Mittenothamnium reptans*: A. portion of secondary stem with branches; B. perichaetium and capsule; C. perigonium; D. inner perichaetial leaf; E. secondary stem leaf; F. branch leaf. (Florschütz 4733).

distal cell end, 30-80 µm long and 2.5-6 µm wide, basal cells shorter, alar cells slightly differentiated in the stem leaves, oval-rectangular, forming a small group, inconspicuous in branch leaves.

Autoicous. Perigonia on branches, leaves oval, gradually acuminate, serrate at apex, to 0.9 mm long. Perichaetia on secondary stems, leaves to 2.5 mm long, ovate-lanceolate, abruptly long-acuminate, acumen serrate, reflexed when dry.

Seta 2-3 cm long, reddish, smooth; capsule inclined, oval-cylindric, 2.5 mm long, operculum conic-rostrate; peristome with the characters of the genus, exostome teeth to 600 µm long, endostome segments perforated, papillose, cilia 1-2, very fragile. Calyptra narrowly cylindric.

D i s t r i b u t i o n : West Indies, Central and tropical South America, tropical Africa.

E c o l o g y : Epiphytic on bark of trees and on logs, occasionally on litter over rocks. Growing preferably in open or low vegetations, usually at higher altitudes.

S e l e c t e d   s p e c i m e n s : Guyana: Upper-Mazaruni district, Pwipwi Mt., Alt. ca. 800 m, Gradstein 5704 (U). Suriname: Emma Mts., Alt. 500 m, Gonggrijp & Stahel 166A (U); Nassau Mts., Alt. ca. 400 m, Geijskes 6 (U); Brownsberg plateau, Alt. ca. 450 m, Florschütz 4733 (U). French Guiana: Mt. de l'Inini, Alt 800 m, Cremers et al. 9040 (CAY, U).

N o t e : This species is distinguished from *Chrysohypnum diminutivum* by the arched, stipitate secondary stems, the cells projecting at the distal ends only and the foliose pseudoparaphyllia. These are also the characters that separate this genus from *Chrysohypnum* (Buck 1984). The distant triangular leaves on the secondary stem form a good additional character.

5. **PHYLLODON** Schimp. in B.S.G., Bryol. Eur. 5: 60. 1851.
   Type: P. truncatulus (C. Müll.) W.R. Buck (Hypnum truncatulum C. Müll.)

Glossadelphus Fleisch., Musci Fl. Buitenzorg 4: 1351. 1923.

Slender plants with creeping stems, subpinnately branched. Leaves oblong to lingulate with obtuse or truncate apex, costa short and double, margin subentire to strongly dentate with bifid teeth in upper part; midleaf cells linear, flexuose, with serially arranged papillae or/and projecting cell ends, alar cells little differentiated.

Seta elongate, smooth; capsule ovoid, inclined, operculum conic; peristome double, exostome teeth not furrowed, transversely striate at base and papillose at tip, endostome with a high basal membrane and keeled segments, cilia present.

N o t e : For nomenclatural changes in this genus see Buck (1987).

D i s t r i b u t i o n : pantropical; only one species in the Guianas.

1. **Phyllodon truncatulus** (C. Müll.) W.R. Buck, Mem. New York Bot. Gard. 45: 519. 1987. – *Hypnum truncatulum* C. Müll., Syn. 2: 263. 1851. – *Glossadelphus truncatulus* (C. Müll.) Fleisch., Musci Fl. Buitenzorg 4: 1352. 1923. Type: Peru, Pöppig s.n. (BM).
– Fig. 174.

Hookeria retusa Wils., nom. nud. Material: Peru, Casapi, Matthews 821 (BM).
Hypnella jamesii Robins., Bryologist 68: 333. 1965, syn. nov. Type: Costa Rica, Monte Verde, James 32 (US).

Light green to dull green plants growing in rather dense mats. Stems long, creeping, defoliate with age, irregularly or subpinnately branched; branches to 1.5 mm long, sparingly divided, prostrate; pseudoparaphyllia foliose. Leaves usually strongly complanate, 0.5-0.8 mm long and 0.25-0.4 mm wide, towards ends of branches diminishing in size; dorsal leaves ovate-oblong, appressed, lateral leaves oblong-lingulate, patent-spreading, concave to conduplicate; costae usually short, but sometimes one or both well developed, extending to 1/4 of leaf length; apex truncate or rounded; margin in upper part coarsely dentate, often with bifid or trifid teeth, in lower part serrulate. Leaf cells thin walled, flexuose-linear, with 1-4 spiny papillae on outer surface, at midleaf 30-60 µm long and 3-5 µm wide, in apex oblong to irregularly isodiametric, to 8 µm wide; basal cells wider, oblong-linear, slightly incrassate, smooth; alar cells hardly differentiated, rectangular or oval, to 20 µm long and 8 µm wide. Papillae slender, irregularly scattered over the cell lumina, frequently at cell ends, often arranged in transverse rows over the leaf, increasing in length towards leaf apex, to 10 µm long, more or less curved.
Synoicous. Perichaetial leaves to 1 mm high, convolute at base, with acute, dentate apex. Seta smooth, reddish, to 2 cm long; capsule ovoid, inclined, operculum conic, rostellate; peristome with the characters of the genus, exostome brown, about 400 µm high, densely striate in basal part; endostome segments broad, keeled, cilia single.

Distribution: Peru, Ecuador, Colombia, Brazil, French Guiana, West Indies, Central America; reported also from Togo, Africa.

Ecology: On tree bases and rocks in moist forest. Collected only twice in French Guiana, not known from Suriname or Guyana.

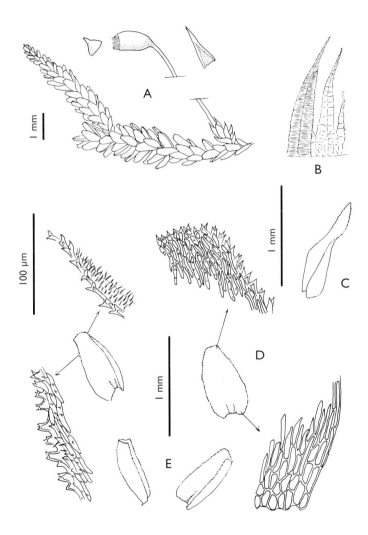

Fig. 174. *Phyllodon truncatulus*: A. stem portion with branch and capsule; B. part of peristome: exostome tooth, endostome segment and cilia; C. perichaetial leaf; D. dorsal leaf; E. lateral leaf. (Montfoort & Ek 509).

S p e c i m e n s e x a m i n e d : French Guiana: Saül, 2 km SW of the village, Alt. 180-210 m, Montfoort & Ek 509 (U); Eaux Claires, 5 km N of Saül, Sentier botanique, Alt. 100-400 m, Florschütz-de Waard 5910 (U).

N o t e s : A peculiar moss, easily recognized by its truncate leaf apex with strongly dentate margins and numerous long, slender papillae. At first sight it could be taken for a species of *Hypnella*, but it is different in the sharply dentate leaf margin and the lack of a well-developed double costa. The latter character however is variable in both genera.
Sterile collections from Costa Rica, described as *Hypnella jamesii* (Robinson 1965), proved to be identical with *Phyllodon truncatulus.*

6. **RHACOPILOPSIS** Ren. & Card., Rev. Bryol. 27: 47. 1900.
Type: R. trinitensis (C. Müll.) Britt. & Dix. (Hypnum trinitense C. Müll.)

Dimorphella (C. Müll.) Ren & Card., Bull. Soc. Roy. Bot. Belgique 41: 101. 1905, nom. illeg. incl. gen. prior.

Plants with creeping stems, irregularly branched, branches complanate-foliate. Branch leaves dimorphous, dorsal leaves asymmetric, ovate, ventral leaves usually narrower, symmetric, costae short and double; leaf cells smooth, oblong-linear, alar cells differentiated.
Dioicous. Seta elongate, smooth; capsule small, subpendulous; peristome double, exostome teeth transversely striolate, endostome segments on a high basal membrane.

N o t e : Until recently this genus consisted of only one species. Buck (1993) concluded that the African genus *Acanthocladiella* is synonymous, adding four species with not or little differentiated ventral leaves to the genus. This makes the genus concept less distinct.

D i s t r i b u t i o n : neotropics and tropical Africa; only one species in the Guianas.

1. **Rhacopilopsis trinitensis** (C. Müll.) Britt. & Dix., J. Bot. 60: 86. 1922. – *Hypnum trinitense* C. Müll., Syn. 2: 284. 1851. – *Ectropothecium trinitense* (C. Müll.) Mitt., J. Linn. Soc. Bot. 12: 86. 1869. Type: Trinidad, Mt. Tocuche, Crüger s.n.(BM).   – Fig. 175.

Dimorphella pechuellii (C. Müll.) Ren. & Card., Bull. Soc. Roy. Bot. Belgique 41: 101. 1905. – *Hypnum pechuellii* C. Müll., Flora 69: 523. 1886. Type: Africa, Kuili River, near Pelle ma Nanga, Pechuël-Lösche s.n.

Slender, pale-green, glossy plants, growing in loose mats. Stems elongate, creeping, irregularly branched, branches elongate, prostrate or pendent; pseudoparaphyllia filamentous or slenderly foliose. Stem leaves erect-spreading or slightly homomallous, ovate, long-acuminate; branch

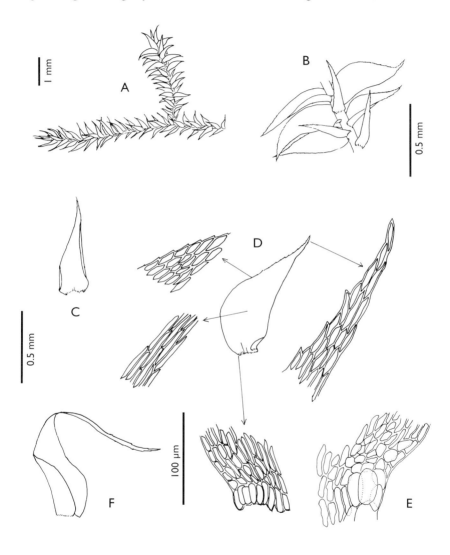

Fig. 175. *Rhacopilopsis trinitensis*: A. stem portion with branch; B. part of branch, ventral view; C. ventral leaf; D. lateral leaf; E. alar cells (coll. Tjon Lim Sang et al. 45); F. inner perichaetial leaf. (A-D, F: Cremers 5131).

leaves distant, dimorphous: dorsal and lateral leaves complanate-spreading, asymmetric to falcate, ovate, long- or short-acuminate, to 1.4 mm long and 0.6 mm wide, ventral leaves narrower, erect, subsymmetric, ovate-lanceolate to narrowly triangular, margins serrulate, sometimes serrate towards apex, costae extending to 1/5 of leaf length, often indistinct; leaf cells more or less incrassate, at midleaf oblong-linear, often flexuose, 30-85 µm long and 4-7 µm wide, in the falcate leaves shorter towards the convex margin, basal cells wider, oval, incrassate, alar cells differentiated, rectangular or quadrate, to 25 µm long and 15 µm wide, partly inflated and hyaline, sometimes forming small auricles.

Dioicous? Perigonia not seen. Perichaetia rare, perichaetial leaves ovate, abruptly long-acuminate, to 1.8 mm long, flexuose, dentate at apex. No capsules seen, description after Bartram (1949) and Crum (1994): seta 10-20 mm long; capsule inclined to subpendulous, about 1 mm long, oblong, contracted to a short neck, operculum conic, 0.5 mm long; peristome teeth brownish, exostome teeth transversely striolate, papillose, endostome segments keeled, on a high basal membrane. The descriptions of the sporophyte by Müller (1851) and Mitten (1869) are based on different components of the Crüger collections according to Dixon (1922).

Distribution: West Indies, Central and tropical South America, tropical Africa.

Ecology: In lowland rainforest, also at higher altitudes; epiphytic on bark of trees and on rotten logs, also in the canopy, occasionally epiphyllous. Not frequent, but probably often overlooked due to the small size.

Selected specimens: Guyana: Mazaruni distr., Mt. Latipu, Alt. 700 m, Maas et al. 2661A (U). Suriname: Brownsberg, Alt. 500 m, Florschütz 4624 (U); Lely mts., Alt. 550-700 m, Lindeman, Stoffers et al. 504A (U). French Guiana: Haut Maroni, Cremers 5131(CAY, U); Eaux Claires, 5 km N of Saül, Alt. 400 m, Florschütz-de Waard 5943 (U).

Notes: The dimorphous branch leaves, seemingly arranged in 4 rows, make this species usually easy to recognize: the falcate leaves on the dorsal side are wide-spreading laterally, giving the branches a strongly complanate appearance; the slender ventral leaves in 2 rows are only visible on the lower side of the branches. In a few collections the ventral leaves are less differentiated (broader and ovate), which goes together with less complanate branches. This makes these specimens sometimes difficult to recognize, but often some branches with the typical leaf arrangement can be discovered in these collections. The shorter leaf cells along the convex (front) margin of the falcate leaves form a useful additional character.

(front) margin of the falcate leaves form a useful additional character.
The alar cells are variably differentiated; usually only the basal cells are slightly inflated, but occasionally a group of alar cells is distinctly inflated, forming small auricles. Dixon (1922) specially studied this character trying to distinguish this neotropical species from the African species *Dimorphella pechuellii*. The inflated alar cells are more frequent in the neotropical material, but he concluded that the high variation made the separation impossible; this made the genus *Dimorphella* synonymous with *Rhacopilopsis*.

7. **VESICULARIA** (C. Müll.) C. Müll., Bot. Jahrb. 23: 330. 1896. – *Hypnum* subsect. *Vesicularia* C. Müll., Syn.2: 233. 1851.
   Type: V. meyeniana (Hampe) Broth. (Hookeria meyeniana Hampe).

Slender to medium-sized plants with creeping, irregularly or pinnately branched stems, branches prostrate. Leaves asymmetric to falcate, broad-ovate to oblong-lanceolate, apex short- to long-acuminate, margin entire or serrulate at apex, costae short and double, often indistinct; leaf cells lax, oval-rhomboidal to elongate-hexagonal, sometimes narrower in the marginal row, alar cells not differentiated.
Autoicous or dioicous. Seta elongate, smooth; capsule horizontal to pendent, ovoid, operculum apiculate to bluntly rostrate; peristome double, exostome teeth lanceolate, striolate in basal part, papillose at apex, endostome with broad, keeled segments and well-developed cilia on a high basal membrane.

D i s t r i b u t i o n : pantropical; only one species in the Guianas.

1. **Vesicularia vesicularis** (Schwaegr.) Broth., Nat. Pfl. Fam. 1(3): 1094. 1908. – *Hypnum vesiculare* Schwaegr., Spec. Musc. Suppl. 2(2): 167. *199*. 1827. Type: Jamaica, Richmond, Reider s.n. (G).   – Fig. 176.

   Vesicularia surinamense (Dozy & Molk.) Broth., Nat. Pfl. Fam. 1(3): 1094. 1908. – *Hypnum surinamense* Dozy & Molk., Prodr. Fl. Bryol. Surin.: 25 (Pl. 14). 1854. Type: Suriname, Splitgerber s.n. (L).

Slender, green to yellowish-green plants growing in flat mats. Creeping stems often strongly flattened, irregularly to pinnately branched, branches short, complanate; pseudoparaphyllia filiform or narrowly lanceolate. Stem leaves homomallous, branch leaves complanate-spreading or homomallous, twisted when dry, broad-ovate to ovate-lanceolate, often falcate, 0.7-1.2 mm long and 0.3-0.8 mm wide, ventral leaves sometimes

460

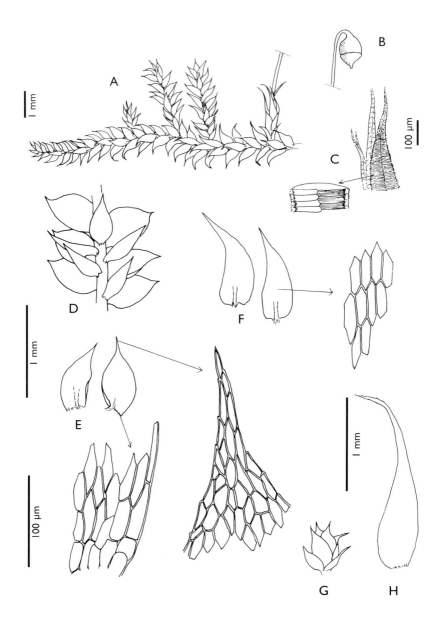

Fig. 176. *Vesicularia vesicularis*: A. stem portion with branches and perichaetium; B. capsule; C. part of peristome with detail of exostome tooth; D. dimorphous leaves, ventral view of branch; E. stem leaves; F. branch leaves; G. perigonium; H. inner perichaetial leaf. (A-C, E-H: Florschütz 4868; D: Florschütz 455).

serrulate in apex, costae extending to 1/5 (occasionally to 1/3) of leaf length, sometimes absent. Midleaf cells thin walled or firm walled, rhomboidal or hexagonal, 30-100 µm long and 10-18 µm wide, in ventral leaves often narrower, basal cells shorter, alar cells not differentiated. Autoicous. Perigonial leaves ovate, short-acuminate, to 0.9 mm long. Perichaetial leaves ovate-lanceolate, gradually long-acuminate, to 2 mm long. Seta smooth, 1-2 cm long; capsule inclined to pendent, ovoid, operculum bluntly short-rostrate; peristome with the characters of the genus, exostome teeth 350-400 µm long, endostome segments minutely papillose, cilia 1 or 2.

D i s t r i b u t i o n : Southern U.S., West Indies, Central and tropical South America.

E c o l o g y : A common moss in the understory of moist forest; epiphytic on tree bases and logs on periodically flooded forest floor, also epilithic or terrestrial near streams. Seldom collected in Guyana.

S e l e c t e d s p e c i m e n s : Guyana: Kanuku Mts., Maipaima, Alt. 160 m, Jansen-Jacobs et al. 993 (U). Suriname: Lely Mts., Alt. 600-650 m, Florschütz 4868 (U); Upper Coppename River, near Bakhuis Mts., Florschütz & Maas 2776 (U); Area of Kabalebo Dam project, marsh forest, Florschütz-de Waard & Zielman 5698 (U). French Guiana: Mt. de l'Inini, Alt. 500 m., Cremers et al. 8901 (CAY, U); Saül, Alt. 180-210 m, Montfoort & Ek 495 (U).

N o t e s : *V. vesicularis* closely resembles *Leucomium* in leaf shape and areolation, but it can be distinguished by the shorter leaf cells and the strongly complanate, often pinnate growth form. For a more detailed comparison see note under the family description of Leucomiaceae (pg. 365)
The variation in leaf shape and areolation of *V. vesicularis* is considerable. Buck (1984), after examining the type specimens of the West Indies, concluded that the morphological variations seem to be controlled by light, humidity and substrate. Although many intermediates between the extreme forms can be found, he proposed 4 varieties, based on the differentiation of dorsal and ventral leaves and on the variation in the dorsal leaves. In the numerous collections of the Guianas these varieties can not be separated. The shape of the dorsal leaves may vary from lanceolate and falcate with acute-acuminate apex to broad-ovate and only slightly asymmetric with short-acuminate apex. In the latter specimens the ventral leaves can be much narrower, causing a certain dimorphism. However, branches with dimorphous leaves may also occur on plants with further undifferentiated leaves.

EXCLUDED SPECIMENS

The Connell & Quelch collections (BM) from Mt. Roraima, labelled as "British Guyana" were taken from the Venezuelan side of the mountain (Steyermark 1981). Collections of *Hypnum amabile* (Mitt.) Hampe and *Ctenidium malacodes* Mitt. were reported for Guyana (Brotherus 1901), but these species were not collected in the Guianas.

A Korthals collection (L), identified as *Taxiphyllum taxirameum* (Mitt.) Fleisch. by Paris in 1907, has the location "Suriname" indicated on the label. This must be a mistake because Korthals never collected in Suriname.

NUMERICAL LIST OF ACCEPTED TAXA

Leucomiaceae

L1. Leucomium Mitt.
      L1-1 L. steerei B.H. Allen & Veling
      L1-2 L. strumosum (Hornsch.) Mitt.

Thuidiaceae

T1. Cyrtohypnum (Hampe) Hampe & Lor.
      T1-1 C. involvens (Hedw.) W.R. Buck & Crum
      T1-2 C. scabrosulum (Mitt.) W.R. Buck & Crum
      T1-3 C. schistocalyx (C. Müll.) W.R. Buck & Crum

T2. Thuidium Schimp.
      T2-1 T. peruvianum Mitt.
      T2-2 T. tomentosum Schimp.

Sematophyllaceae

S1. Acroporium Mitt.
      S1-1 A. pungens (Hedw.) Broth.

S2. Meiothecium Mitt.
      S2-1 M. boryanum (C. Müll.) Mitt.
      S2-2 M. commutatum (C. Müll.) Broth.

S3. Potamium Mitt.
      S3-1 P. deceptivum Mitt.
      S3-2 P. vulpinum (Mont.) Mitt.

S4. Pterogonidium C.Müll.
      S4-1 P. pulchellum (Hook.) C. Müll.

S5. Sematophyllum Mitt.
      S5-1 S. cochleatum (Broth.) Broth.
      S5-2 S. galipense (C. Müll.) Mitt.
      S5-3 S. lonchophyllum (Mont.) J. Florsch.
      S5-4 S. pacimoniense (Mitt.) J. Florsch.
      S5-5 S. subpinnatum (Brid.) Britt.
      S5-6 S. subsimplex (Hedw.) Mitt.

S6. Taxithelium Spruce ex Mitt.
      S6-1 T. concavum (Hook.) Spruce
      S6-2 T. planum (Brid.) Mitt.
      S6-3 T. pluripunctatum (Ren. & Card.) W.R. Buck

S7. Trichosteleum Mitt.
      S7-1 T. bolivarense Robins.
      S7-2a T. hornschuchii (Hampe) Jaeg. var. hornschuchii
      S7-2b T. hornschuchii (Hampe) Jaeg. var. subglabrum J. Florsch.
      S7-3 T. intricatum (Thér.) J. Florsch.
      S7-4 T. papillosum (Hornsch.) Jaeg.

S8. Wijkia Crum
      S8-1 W. costaricensis (Bartr. & Dix.) Crum

Hypnaceae

H1. Chrysohypnum Hampe
      H1-1 C. diminutivum (Hampe) W.R. Buck

H2. Ectropothecium Mitt.
      H2-1 E. leptochaeton (Schwaegr.) W.R. Buck

H3. Isopterygium Mitt.
      H3-1 I. subbrevisetum (Hampe) Broth.
      H3-2 I. tenerum (Sw.) Mitt.

H4. Mittenothamnium Henn.
      H4-1 M. reptans (Hedw.) Card.

H5. Phyllodon Schimp.
      H5-1 P. truncatulus (C. Müll.) W.R. Buck

H6. Rhacopilopsis Ren. & Card.
      H6-1 R. trinitensis (C. Müll.) Britt. & Dix.

H7. Vesicularia (C. Müll.) C. Müll.
      H7-1 V. vesicularis (Schwaegr.) Broth.

COLLECTIONS STUDIED

GUYANA

Abraham, A.A., 139 (S5-6); 265 (S5-6) – (BM)

Appun, C.F., 404 (S1-1); 725 (S5-6); 815 (S5-6) – (BM); 820 (S6-2) – (NY); 660 (S6-2); 810 (S6-2); 901 (S5-6); s.n. (T2-1) (H6-1) (S5-6) – (BM)

Aptroot, A., 17051 (S5-6); 17053 (S5-5); 17062 (S5-5); 17067 (S7-4); 17094 (S1-1); 18396 (T2-2); 18924B (S7-4); 19326 (S5-6) – (U)

Bartlett, A.W., 8049 (S7-4); 8086 (S1-1); 8090 (S1-1); 8279 (S3-2); 8635 (S1-1) – (BRG)

Biesmeijer, K. & B. Bleij, 8 (L1-2) – (U)

Boom, B. et al., 7148 (S3-1) type of *Maguireella vulpina*; 7247 (S5-5); 7249 (S7-1); 7495 (S1-1); 7656 (S1-1); 9129A (S7-4) – (NY)

Cornelissen, J.H.C. & H. ter Steege, 12 (L1-2); 18 (S5-6); 44 (S5-6); 51 (S7-4); 56 (S6-2); 71 (S1-1); 72 (S5-6); 93 (S7-2a); 95 (S7-4); 106 (L1-2); 141 (S5-6); 164 (S7-4); 168 (S5-6); 179 (S6-2); 193 (L1-2); 198 (S7-2a); 207 (S7-2b); 208 (S7-2b) type of *Trichosteleum horn-schuchii* var. *subglabrum*; 952 (S5-6); 960 (S5-6); 961 (S1-1); 964 (S7-3); 966 (S5-5); 968A (S2-1); 970A (S2-1); 988 (S5-6); 995-997 (S1-1); 1021 (S5-6); 1024 (S5-6); 1073 (S2-1); 1075 (S2-1) – (U)

Ek, R.C. et al., 964A (S6-3); 1110 (S7-4) – (U)

Florschütz-de Waard, J., 6022 (S7-2a); 6023 (H3-2); 6024 (S6-1); 6033 (S6-1); 6037 (S5-6); 6038 (S7-2a); 6040 (S6-1); 6045 (S5-6); 6046 (S7-2a); 6048 (S3-2); 6049 (S5-3); 6052 (S7-2a); 6053 (S5-5); 6063 (S5-5); 6065 (S5-6); 6070 (S6-2); 6073 (S7-2b); 6075 (S7-4); 6076 (S7-2a); 6077 (L1-2); 6080 (S7-2b); 6081 (S7-2b); 6082 (H3-2); 6083 (S5-6); 6088 (S6-3); 6089 (S5-6); 6092 (S1-1); 6095 (S6-2); 6096 (S5-5); 6099 (S5-6); 6100 (L1-2); 6101 (S5-5); 6104 (S5-5); 6105 (S7-2a); 6113 (S4-1); 6114 (S6-2); 6115 (S5-5); 6120 (S7-4); 6125 (S5-5); 6126 (H3-2); 6127 (L1-2); 6128 (T1-2); 6133 (H3-2); 6134 (S5-6); 6135 (S6-3); 6136 (S7-4); 6138 (H3-2); 6139 (S7-2b); 6140 (S3-2); 6141 (H3-2) – (U)

Gradstein, S.R., 4711 (S5-5); 4712 (S5-6); 4713 (S6-2); 4723 (S5-5); 4723A (S3-1); 4782 (H3-2); 4783 (S3-1); 4809 (S5-6); 4810 (S7-4); 4817 (S1-1); 4846 (S7-2a); 4846A (S7-1); 4847 (S5-6); 4854 (H4-1); 4865 (L1-1); 4865A (S7-1); 4882 (T2-2); 4926 (S6-2); 4932 (H2-1); 4944 (H1-1); 4947 (H2-1); 4958 (S5-6); 4960 (H2-1); 4962 (T2-2); 4964 (S3-2); 4970A (S7-2a); 4995 (S1-1); 5025 (S5-5); 5049 (S3-2); 5088 (S5-6); 5103 (S1-1); 5113 (S7-1); 5159 (S7-4); 5163 (S1-1); 5277 (S1-1); 5278 (T2-2); 5286 (L1-1); 5316 (S7-4); 5317 (T2-2); 5398 (S1-1); 5421 (L1-1); 5525 (L1-2); 5529

(S6-2); 5534A (S6-3); 5557 (L1-2); 5564 (H3-2); 5568 (H3-2); 5583 (S6-2); 5651 (S1-1); 5664 (H3-2); 5678 (S7-1); 5690 (S7-1); 5694 (S1-1); 5704 (H4-1); 5715 (H4-1); 5725 (S5-6); 5733 (H7-1); 5736 (S7-4); 5736 (S7-2a) – (U)

Graham, E.B., E4 (S6-2); 99 (S6-2); 136 (S5-6); 164 (S5-5); 185 (S5-5); 201 (S5-6); 208 (S6-2); 283 (S5-6); 347 (S6-2); 350 (S6-1); 372 (S6-2) – (NY)

Harrison, S.G., 586A (S6-3) – (BM)

Hartley, J., 1406 (S7-4); 1423 (S7-4); 1425 (S6-2); 1437 (S5-6) – (NY)

Hofmann, B., 2948 (S5-2) – (US)

Howes, P.S., s.n. (S5-6) – (NY)

Jansen-Jacobs, M.J. et al.,225 (S6-2); 225B (S7-2a); 647 (T1-1); 993 (H7-1); 1239 (T1-1); 1636 (S5-6); 1650 (S1-1) – (U)

Jenman, G.S., 7856 (S6-2); 7862 (T2-2) – (BRG)

Kennedy, H. et al., 4742 (S7-4); 4742A (S6-2); 4775 (S5-6); 4775A (S1-1) – (US)

Leng, H., 4 (H6-1); 8 (S3-2); 365B (S5-6) – (NY)

Linder, D.H., 208 (S5-6); 476 (S6-2) – (NY)

Maas, P.J.M. et al., 2504 (S3-1); 2505 (S6-2); 2593 (S5-6); 2661A (H6-1); 3850 (T1-2); 3859 (H1-1); 3860 (H1-1); 3996 (S6-2); 4144 (H2-1); 4479 (S5-6); 4486 (S5-6); 5588 (S5-6); 5876 (S6-2); 5876A (L1-2); 7479 (S1-1); 7480 (S1-1) – (U)

Noël, E.F., 95 (H4-1); 95A (S6-2); 96A (S5-6) – (BM, NY)

Parker, Ch.S., s.n. (H3-2); s.n. (S4-1); s.n. (S1-1); s.n. (S6-2); s.n. (S5-6) – (NY); hb. Hooker 3238 (S5-6) – (BM)

Pipoly, J.J., 8863 (S7-4); 8870 (10,4); 8908 (S7-4); 8999 (S6-2); 9286 (S6-2); 9304 (S6-2); 10606A (S1-1); 10610A (S1-1) – (NY)

Quelch, Cl. J., hb. Levier 1265 (H3-2) type of *Plagiothecium radicisetum* – (NY); hb. Levier 1267 (L1-2) type of *Leucomium guianense* – (BM); hb. Levier 1269 (S5-6); hb. Levier 1270 (S6-2); s.n. (T1-2) type of *Thuidium verrucipes* – (BM); hb. Levier 1272 (S7-2a) type of *Aptychus micropyxis* – (BM); hb. Levier 1286 (S5-5) type of *Aptychus grammicarpus* – (BM); hb. Levier 1288 (S6-1) type of *Taxithelium quelchii* – (BM); hb. Levier 1289 (S1-1) – (BM); hb. Levier 1293 (T1-2) type of *Thuidium verrucipes*; s.n. (S3-1) type of *Potamium leucodontaceum* – (BM); s.n. (T1-2); s.n. (S5-5); s.n. (S7-4); s.n. (S6-1); s.n. (S5-6); s.n. (S7-2b) – (BM)

Richards, P.W., 89 (L1-2); 105 (S1-1); 110 (L1-2); 164 (S5-6); 195 (S7-4); 207 (S5-6); 241 (H6-1); 271 (S5-6); 360 (S3-2); 371 (S6-2); 377 (H6-1); 377A (S7-3); 406 (S7-2b); 425 (S6-2); 427 (S5-6); 441 (S2-1) ; 479 (S7-4); 482 (S6-2); 497 (L1-2); 501 (S6-2); 538 (S5-5); 745 (S7-4); 748 (S7-2b); 768 (S7-2b); 791 (S7-4); 835 (S6-1); 847 (S3-2); 850 (S6-1) – (BM,NY)

Robinson, H.E., 850022 (S1-1); 850035 (H2-1); 850075 (S5-6); 850184 (S5-5) – (NY,U)

Schomburgk, R., 1608 (S5-6) – (BM)

Smith, A.C., 2107 (S6-2); 2383 (S6-2); 2497 (H1-1); 2498 (S6-2); 2520 (S7-3); 2525 (S5-6); 2546 (S6-2); 2549 (S6-1); 2565 (S3-2); 2574 (S6-2); 2604 (S5-6); 2636 (S1-1); 2644p.p. (S3-2); 2644p.p. (S6-1); 2649 (S6-1); 2655 (S6-1); 2760 (S1-1); 2854 (H7-1); 2878p.p. (S6-2); 2895p.p. (S6-2); 2949A (S1-1); 2975A (S7-4); 2976 (S5-6); 3088 (S5-6); 3093 (H1-1); 3119 (S5-2); 3197 (H1-1); 3220 (S5-2); 3357 (S5-6); 3415 (S5-6); 3618 (H6-1); 3629 (S7-4); 3630 (S5-6); 3632A (S5-2); 3635 (S5-2) type of *Rhaphidostichum guianense*; 3672A (S5-2) – (NY, U)

Tutin, T.G., 199 (S5-6); 257 (S6-2); 285 (S5-6); 369A (S6-2); 538 (T2-2); 650 (S7-2b); s.n. (S7-2b) – (BM)

SURINAME

Anakam, H.F., s.n. (T1-1) – (BM); s.n. (S5-5) – (NY)

Aptroot, A., 17055 (S5-5) – (U)

Arnoldo, Fr., 3494 (S5-6); 3504 (S5-6); 3506 (H2-1) – (U)

Beek Vollenhoven, van, s.n. (S4-1) – (L)

Bekker, J.M., 1038B (H6-1); 1051 (H6-1); 1202A (S7-4); 1286B (S7-4); 1306A (S7-4); 1347B (S6-1); 1669 (H7-1) – (U)

Benjamins, H.D., s.n. (S5-6); s.n. (S6-1); s.n. (S6-2) – (L)

Boerboom, T., 10 (S7-2a); 24 (S7-4); 41 (H3-2); 44 (S7-2a); 58 (H2-1); 63 (S5-6); 81 (H6-1); 90 (S7-4); 94 (H2-1); 100 (H3-2) – (U)

Boon, H.A., 1127A (S6-1) – (U)

Donselaar, J. van, 1562 (H7-1); 1563 (S6-1); 1564 (S4-1); 1565 (S5-3); 1572 (S6-1); 1576 (S6-2); 1580 (S5-3); 1585 (S5-5); 1592 (S5-5); 1614 (S1-1); 1615 (S6-3); 1617 (S5-6); 1618 (S7-4); 2460 (S1-1); 2586 (S7-2a); 2759 (S1-1); 2788 (S5-6); 2791 (S1-1); 3097 (L1-2); 3101 (S6-2); 3283A (T1-2); 3296 (H7-1); 3297 (S6-2); 3722 (S5-1); 3723 (S6-1); 3724 (S5-1); 3726 (S5-6) – (U)

Florschütz, P.A., 125 (S6-2); 129 (S4-1); 133 (S4-1); 135 (S5-5); 139 (S4-1); 142 (H1-1); 142A (H4-1); 180 (T1-2); 181 (S7-2a); 196 (S5-1); 197 (H7-1); 205 (S5-1); 214 (S5-1); 218 (S5-1); 222 (S5-1); 229 (S6-1); 230 (S5-5); 231 (H7-1); 254 (S5-5); 264 (S7-2a); 265 (S6-1); 268 (S5-1); 284 (L1-2); 285 (T1-2); 291 (S6-2); 294 (S7-4); 298 (H7-1); 301 (S7-3); 307 (H7-1); 308 (H3-2); 320 (H2-1); 321 (H3-2); 323 (H2-1); 325 (T1-2); 330 (T1-2); 338A (S6-3); 362 (H2-1); 363 (S7-4); 368 (S5-5); 369 (L1-2); 370 (H7-1); 383 (S5-6); 384 (S6-2); 387 (H2-1); 389 (S7-4); 392 (T1-2); 393 (S1-1); 397 (H7-1); 406 (L1-2); 413 (H2-1); 416 (L1-2); 431 (H7-1); 433 (S7-3); 435 (H3-2); 455 (H7-1); 457 (L1-

2); 462 (H2-1); 467 (H2-1); 495 (S6-1); 505 (S5-5); 527 (H3-2); 528 (S3-1); 529 (S5-5); 530 (S7-2a); 539 (S4-1); 563 (S4-1); 567 (H7-1); 576 (S5-5); 578 (S6-2); 589 (H7-1); 591 (H1-1); 594 (T1-3); 606 (S4-1); 607 (H3-2); 611 (H3-2); 655 (S5-6); 663 (S7-4); 669 (H3-2); 672 (S5-6); 673 (S7-2a); 678 (S7-4); 713 (S5-6); 728 (S7-4); 774 (S6-2); 779 (S7-3); 787 (S7-2b); 837 (H3-2); 852 (S5-6); 859 (S7-4); 861 (S1-1); 861A (H6-1); 871 (S5-5); 890 (S5-5); 976 (S5-5); 977 (S6-2); 1045 (S5-5); 1046 (S4-1); 1079 (S4-1); 1105 (S6-1); 1129 (S7-2a); 1138 (H7-1); 1139 (S6-2); 1140 (S5-5); 1141 (S5-6); 1146 (S7-2a); 1148 (S6-1); 1 170 (S5-5); 1171 (S6-1); 1176 (S6-1); 1189 (S7-2a); 1190 (S5-6); 1191 (S6-1); 1193 (S5-5); 1200 (S7-2b); 1203 (L1-2); 1205 (H2-1); 1218 (S6-1); 1220 (S7-2b); 1221 (S6-1); 1222 (S3-2); 1224 (S5-5); 1226 (S3-2);1232 (L1-2); 1254 (S7-4); 1259 (T1-2); 1263 (S5-6); 1264 (L1-2); 1265 (T1-2); 1266 (H2-1); 1269 (L1-2); 1315 (S5-6); 1316 (S2-1); 1332 (S6-2); 1333 (L1-2); 1338 (H7-1); 1342 (L1-2); 1358 (T2-2); 1360 (T1-2);1361 (H7-1); 1362 (S5-5); 1365 (L1-2); 1373 (S1-1); 1418 (S6-1); 1419 (H6-1); 1420 (H6-1); 1423 (S5-6); 1424 (S7-4); 1429 (T2-2); 1432 (L1-2); 1434 (S3-2); 1473 (S5-6); 1477 (S5-2); 1484 (H7-1); 1484A (T1-2); 1491 (S5-5); 1493 (S5-5); 1496 (S1-1); 1504 (S8-1); 1517 (T1-2); 1535 (S5-1); 1539 (S5-5); 1540 (H3-2); 1543 (S7-4); 1558 (H7-1); 1581 (S5-5); 1583 (S5-6); 1586 (S2-1); 1594 (S5-5); 1619 (H7-1); 1622 (S7-4); 1623 (S6-1); 1624 (S6-1); 1639 (S6-2); 1640 (T2-2); 1642 (T2-2); 1650 (S5-2); 1660 (S6-1); 1671 (S6-1); 1676 (S2-1); 1697 (H7-1); 1702 (H7-1); 1715 (H7-1); 1717 (S4-1); 1723 (S5-5); 1751 (S5-6); 1759 (T1-2); 1765 (H7-1); 1765 B (H4-1); 1766 (T1-2); 1770 (S6-2); 1780 (S6-3); 1782 (S7-4); 1796 (S1-1); 1804 (S6-2); 1805 (S6-3); 1838 (S7-4); 1839 (S5-6); 1843 (S6-3); 1896 (S6-2); 1903 (S4-1); 1968 (S5-6); 1971 (S7-2a); 1974 (S6-2); 1976A (T1-2); 1977 (H3-2); 2026 (S7-4); 2026A (S6-3); 2027 (S5-6); 2037 (L1-2); 2060 (S4-1); 2062 (S2-1); 2064 (S7-2a); 2066 (S3-1); 2067 (S4-1); 2069 (S5-5); 2070 (S3-1); 2071 (S4-1); 2074 (S5-5); 2075 (H3-2); 2077 (S3-1); 2088 (S7-2a); 2091 (S4-1); 2092 (S6-1); 2093 (S7-2a); 2094 (S6-1); 2096 (S3-2); 2097 (S3-2); 2098 (S6-1); 2099 (S3-1); 2114 (H7-1); 2118 (S6-1); 2123 (S7-4); 2135 (H7-1); 2137 (T1-3); 2140 (S7-2a); 2142 (S6-1); 2151 (H7-1); 2154 (S5-6); 2156 (L1-2); 2157 (S6-3); 2187 (L1-2); 2193 (H3-2); 2201 (T1-1); 2201A (H7-1); 2202 (L1-2); 2203 (H3-2); 2206 (H7-1); 2221 (T1-2); 2228 (S7-4); 2229 (S5-6); 2242 (S6-1); 2244 (S7-2a); 2258 (S7-2a); 2259 (L1-2); 2270 (H7-1); 2275 (S6-2); 2278 (T1-2); 2287 (L1-2); 4526 (S5-6); 4532 (S6-

1); 4538 (T1-3); 4542 (S6-1); 4547 (S6-1); 4547A (S5-5); 4548 (S3-2); 4549 (S6-1); 4550 (T1-2); 4554 (H7-1); 4555 (S6-2); 4562 (T1-2); 4563 (S5-5); 4581 (S7-4); 4584 (S4-1); 4585 (S4-1); 4593 (H3-2); 4593A (S7-2a); 4594 (S6-2); 4595 (H3-2); 4608 (S4-1); 4612 (S5-5); 4617 (H7-1); 4620 B (H6-1); 4623 (S5-5); 4624 (H6-1); 4625 (H2-1); 4639 (S5-5); 4640 (T2-2); 4646 (S7-4); 4647 (H7-1); 4676 (S7-1); 4690 A (H2-1); 4695 (S5-6); 4708 (H2-1); 4709 (H7-1); 4710 (S1-1); 4713 (T1-2); 4714 (H7-1); 4727 (H2-1); 4733 (H4-1); 4736 (H2-1); 4739 (H7-1); 4747 (H4-1); 4748A (T2-2); 4749 (H7-1); 4750A (H4-1); 4766 (S7-4); 4767 (S5-6); 4772 (S5-6); 4768 (T2-2); 4779 (S5-5); 4779A (S4-1); 4783 (S3-2); 4784 (S5-4); 4796B (S2-1); 4798 (S7-1); 4848 (S8-1); 4801 (S7-4); 4802 (H3-2); 4807A (H7-1); 4809 (S5-5); 4810 (H2-1); 4812 (S6-2); 4814 (T2-2); 4817 (S5-5); 4824 (S5-6); 4837A (T2-2); 4847 (S1-1); 4849A (H6-1); 4856 (T2-2); 4858 (S5-6); 4860 (H2-1); 4862 (S5-5); 4865 (L1-2); 4868 (H7-1); 4895 (S6-2); 4896 (S1-1) – (U)

Florschütz, P.A. & P.J.M. Maas, 2310A (L1-2); 2311 (H7-1); 2340 (S4-1); 2352 (H7-1); 2383 (L1-2); 2386 (T1-2); 2389 (T1-2); 2454 (T2-2); 2606 (T2-2); 2607 (S6-2); 2647 (H7-1); 2683 (S4-1); 2689 (T1-2); 2698 (S3-2); 2742 (S5-5); 2774 (L1-2); 2776 (H7-1); 2777 (T1-2); 2848 (L1-2); 2872 (S6-2); 2874 (T2-2); 2884 (H6-1); 2919 (S1-1); 2967 (S1-1); 2970 (S5-6); 2993 (S7-1); 3097A (H4-1); 3118 (H4-1); 3157B (S7-4); 3159 (S3-2) – (U)

Florschütz-de Waard, J., 4887 (H6-1); 4889 (L1-2); 4892 (T2-2); 4897 (S5-5); 4899 (H2-1); 4900 (T2-2); 4901 (H7-1); 4903 (H2-1); 4905 (T2-2); 4906 (H4-1); 4917 (T1-1); 4926 (S4-1); 4930 (S4-1); 4934 (S5-5) – (U)

Florschütz-de Waard, J & R. Zielman, 5009 (S4-1); 5010 (S5-5); 5011 (S6-2); 5012 (S5-5); 5013 (S1-1); 5014 (S5-6); 5017 (H3-2); 5019 (T2-2); 5022 (S5-5); 5023 (S7-4); 5025A (T2-2); 5029 (H6-1); 5030 (S1-1); 5033 (L1-2); 5036A (H3-1); 5040 (T2-2); 5045 (H6-1); 5049 (H2-1); 5050 (T2-2); 5051 (L1-2); 5056 (H2-1); 5058 (T2-2); 5059 (H2-1); 5068 (L1-2); 5069 (S6-1); 5076 (S7-4); 5081 (S6-3); 5083 (L1-2); 5095 (S5-6); 5101 (S7-4); 5112 (S5-6); 5115 (S7-3); 5142 (S6-2); 5144 (L1-2); 5151 (L1-2); 5162 (S7-2a); 5171 (S6-3); 5178 (S5-6); 5183 (S6-1); 5191 (S5-5); 5194 (S7-2a); 5195 (S6-2); 5197 (L1-2); 5200 (S6-2); 5201 (S6-3); 5203 (L1-2); 5211 (S5-6); 5217 (L1-2); 5222 (S5-6); 5223 (S7-4); 5227 (S6-2); 5229 (S7-4); 5229A (H2-1); 5236 (T1-2); 5237 (L1-2); 5248 (S7-4); 5259 (S5-6); 5263 (S7-4); 5264 (S1-1); 5271 (S6-1); 5276 (S3-2); 5277 (S6-1); 5283 (S5-6); 5303 (S7-4);

5304 (L1-2); 5318 (L1-2); 5328 (H7-1); 5330 (H7-1); 5333 (L1-2); 5345 (H7-1); 5346 (H7-1); 5347 (H2-1); 5349 (S7-2a); 5353 (T1-2); 5354 (H7-1); 5357 (S6-2); 5364 (T1-2); 5372 (S7-4); 5376 (S5-5); 5377 (L1-2); 5382 (T1-2); 5390 (S5-6); 5394 (S7-3); 5398 (S7-2a); 5401 (S7-4); 5410 (S7-3); 5414 (L1-2); 5431 (S7-3); 5434 (S7-4); 5447 (S5-5); 5451 (S7-4); 5455 (S5-6); 5457 (S7-3); 5459 (4.); 5465 (S7-4); 5470 (L1-2); 5477 (S7-2a); 5478 (S7-2a); 5480 (H1-1); 5486 (T1-2); 5487 (S5-6); 5489 (S5-5); 5495 (S6-2); 5496 (H3-2); 5507 (T1-2); 5514 (S5-6); 5520 (S5-5); 5540 (L1-2); 5549 (S7-4); 5564 (T1-2); 5569 (L1-2); 5573 (S6-3); 5577 (S7-2a); 5583 (L1-2); 5587 (S5-6); 5591 (S5-5); 5592 (H1-1); 5605 (S7-4); 5611 (S5-6); 5616 (S5-5); 5624 (S6-2); 5636 (T1-2); 5640 (S7-4); 5641 (H7-1); 5642 (H7-1); 5643 (L1-2); 5656 (S5-6); 5666 (H7-1); 5667 (L1-2); 5668 (S2-1); 5670 (T1-2); 5674 (S5-6); 5675 (H2-1); 5679 (L1-2); 5687 (H3-2); 5688 (S7-2a); 5691 (H7-1); 5694 (L1-2); 5698 (H7-1); 5709 (L1-2); 5711 (H7-1); 5712 (H3-2); 5713 (H7-1); 5725 (H7-1); 5726 (T1-2); 5729 (S3-2); 5730 (S6-1); 5732 (S5-5); 5735 (H1-1); 5740 (L1-2); 5741 (H7-1); 5755 (H7-1); 5756 (H7-1); 5757 (L1-2); 5768 (H7-1); 5772 (L1-2); 5775 (T1-2); 5776 (H7-1); 5777 (T1-3); 5783 (S6-2); 5786 (S5-5); 5788 (S2-1); 5798 (S6-1); 5803 (S5-5) – (U)

Focke, H.C., 14 (S6-2); 17 (S4-1); 1190 (S1-1); 1192p.p. (S5-6); 1192p.p (S7-4); 1192 p.p. (H3-2); 1338 (S1-1)p.p.; 1338p.p. (S7-4); 1388p.p. (H3-1); 1388p.p. (S4-1) – (L,U); s.n. (S6-2) – (BM, NY)

Geijskes, D.C., 4 (S5-6); 4A (S3-2); 6 (H3-2); 6E (T2-2); 6F-a (H7-1); 10 (S6-1); 12 (S3-2); 13B (S4-1); 15A (S6-1); 17 (S6-1); 18 (S6-2); 20 (S3-2); 21 (S6-1); 26 (S5-5); 29 (S6-2); 29A (H1-1); 30 (S1-1); 32A (T1-2); 35 (S5-6); 37A (L1-2); 40 (S4-1); 42A (S6-2); 42B (L1-2); 50 (S5-3); 67 (H4-1); 170 (S5-6); 195C (S1-1); 196 (S4-1); 1005 (S1-1); s.n (S4-1); s.n. (S7-2a); s.n. (S7-2b); s.n. (S7-3); s.n. (S7-4); s.n. (H1-1); s.n. (H2-1); s.n. (H3-2); s.n. (H4-1) – (U)

Gonggrijp, J.W. & G. Stahel, 166 (T2-2); 166A (H4-1) – (U)

Gradstein, S.R., 4607 (S6-2); 4618 (S6-2); 4623 (S4-1); 4630 (S5-6); 4670 (H3-1); 4689 (S5-5) – (U)

Heyligers, P.C., 352 (S5-6); 526 (S5-6); 653 (S5-6) – (U)

Houtkoper, H., s.n. (S2-1); s.n. (S5-5); s.n. (S6-2) – (L)

Hostmann, F.W.R., 184 pp. (H7-1) – (BM); s.n. (S1-1); s.n. (H3-2) – (L)

Hulk, J.F., 229A (S5-5) – (U)

Jonker, F.P. et al., 397 (S6-1); 612 (S6-2); 716 (T2-2); 721 (S7-3); 721A (L1-2); 752 (T2-2); 799 (T2-2); 800 (L1-2); 828 (S7-4); 828C (L1-2); 917 (S1-1); 924 (T2-2); 929 (S5-6); 962 (S1-1); 973 B (H6-1); 1117 (T2-2); 1125 (T2-2); 1157 (H7-1); 1346 (H7-1) – (U)

Kappler, A., s.n. (S6-2) – (L)

Kegel, H., 66 (S2-1); 508 (S2-1); 509 (S5-5); 513 (S5-5) type of *Sematophyllum kegelianum*; 514 (T1-2); 517 (S4-1); 518 (S4-1) type of *Hypnum microtheca*; 740 (S5-5); 990 (L1-2) type of *Acosta cuspidata* ; 1199 (S1-1); 1408 (S6-2); 1409 (S5-6); s.n. (H7-1); s.n. (S4-1); s.n. (S5-5) – (GOET)

Kramer, K.U. & W.A.H. Hekking, 2890 (S7-2b); 2890C (H3-2); 3086A (S5-6); 3318 (S5-6) – (U)

Lanjouw, J., 348A (H3-2); 378 (S5-6); 486 (L1-2); 741 (S6-2); 753 (S6-1) – (U)

Lanjouw, J & J.C. Lindeman 520 (S5-6); 524 (S5-6); 619 (S7-2b); 840 (S5-6); 891 (S6-2); 892 (S5-6); 1218 (S5-3); 1251 (S5-5); 1405 (H3-2); 1548 (H3-2); 1750A (H3-2); 1757 (S5-6); 1770 (S6-2); 1998 (S6-1); 2000 (S6-1); 2015 (S5-5); 2162 (S1-1); 2241 (S7-3); 2294 (S7-4); 2355 (T2-2); 2605 (S1-1); 2626 (H7-1); 2648 (L1-2); 2650 (H2-1); 2669 (T2-2); 2672 (S1-1); 2704 (S6-3); 2780 (S1-1); 2843 (S1-1); 2899 (S7-4); 2953 (S5-6); 2794A (H1-1); 3236 (H7-1); 3355 (S7-4); 3380 (S7-4); 3931 (S6-2) – (U)

Lindeman, J.C., 3927 (S1-1); 3929 (S7-4); 3932 (S5-6); 4458A (T1-2); 4458 C (H7-1); 5696 (S5-3); 5837 (S7-4); 5990 (S5-6); 6007 (S5-6); 6426 (S6-1); 6427B (H2-1); 6932 (L1-2); 7037 (S1-1); 7045A (H2-1); 7046 (S1-1); 7050 (T2-2); 7057 (L1-2); 7060 (H2-1); 7064 (S5-6); 7068 (S6-2); 7077 (T1-2); 7086 (S6-1); 7086A (S7-4); 7088 (S6-1); 7105 (S5-6); LBB 12101 (H4-1) – (U)

Lindeman, J.C. & A.L. Stoffers et al., 47 (S7-4); 164 (T2-2); 167 (S1-1); 218C (T2-2); 219 (S1-1); 248 (S5-6); 255 (T2-2); 326A (T2-2); 344A (T2-2); 504A (H6-1); 582 (T2-2); 754A (T2-2); 764 (S5-6); 839 (T2-2); 840 (S5-3) – (U)

Looy, C.H. van, 4 (H3-2); 6 (S5-5); 7 (S5-5); 18 (S5-3); 20 (S4-1); 21 (S6-1); 27 (S3-1); 43 (T1-3) – (U)

Maas, P.J.M., 3264 (S6-2); 3266 (S5-6); 3361A (S6-3); 3363 (T2-2); 3364 (S6-2); 3370 (S6-2); 3379 (S5-6); BW 10757C (S7-2b) – (U)

Maguire, B. et al., 24033 (S5-6); 24339 (H6-1); 24433 (T2-2); 24434 (T2-2); 24496 (L1-2); 24897 (S7-2a); 45940A (S5-6); 46122A (S5-6) – (BM, NY, U)

Mennega, A.M.W., 119 (S6-1); 121 (H7-1); 556 (H6-1); 563 (S6-2); 576 (S6-2); s.n. (S5-5) – (U)

Moonen, J.M., s.n. (S5-4) – (U)

Pulle, A.A., 109 (S5-6); 130A (S7-4); – (U)

Rombouts, H.E.,124A (S5-6); 124B (S6-2); 130A (S6-1); s.n. (T2-2); s.n. (S5-5) – (U)

Schulz, J.P. et al., 7208A (T1-2); 7209 (H7-1); 8055 (T2-2); LBB 10014A (S6-1); LBB 10575 (T1-2); LBB 10578 (S5-6); 10675 (H1-1); 10677 (H3-2) – (U)

Splitgerber, F.L., Hb. Ac. Nat. Sc. Phyladelphia nr-12 (H7-1) type of *Vesicularia surinamense* – (L); s.n. (S5-5); s.n. (S5-6). – (PC)

Stahel, G., 473 (BW 7200) (H7-1); 522 (BW7149) (T2-2); c.p-3 (S5-5); c.p-4 (S6-2) – (U)

Stahel, G. & J.W. Gonggrijp, BW 523C (S1-1);BW 2961A (H3-2) – (U)

Suringar, W.F.R., s.n. (S5-5); s.n. (S5-6); s.n. (S6-1); s.n. (S6-2); s.n. (S7-4); s.n. (S7-2a) as *Trichosteleum suringari* Broth.& Par, nom. herb. – (L)

Sypesteyn, C.A. van, s.n. (L1-2); s.n. (S1-1) – (L)

Terpstra, W.J., 9B (S6-2) – (L)

Teunissen, P.A. & J.Th. Wildschut, LBB 11931 (H3-2) – (U)

Tjon Lim Sang, R. & I.H.M. van der Wiel, 14 (H6-1); 18 (H7-1); 19 (S6-3); 30 (S7-2a); 32 (H7-1); 34 (S6-3); 37 (H6-1); 38 (S1-1); 40 (S6-3); 45 (H6-1); 46 (H7-1); 47 (S7-4); 49 (H3-2); 50 (S7-4); 52 (S6-3); 53 (S6-2) – (U)

Tresling, J., B 189 (H6-1); 193B (S6-2); 193C (H2-1); 288 (S3-2); 302A (S3-2); 384A (S3-2); 384B (S5-5) – (U)

Tulleken, L., 6B (S5-1); 174 (H7-1) – (L)

Wagenaar Hummelinck, P., s.n. (H7-1); s.n. (S7-2a); s.n. (S6-1) – (U)

Weigelt, C., s.n. (S4-1) – (M); s.n. (S5-6) – (BM, L, M); s.n. (S7-4) type of *Hypnum spirale* – (BM); s.n. (S2-1) syntype of *Meiothecium boryanum* – (BM, L, NY)

Wessels Boer, J.G., 180A (H4-1); 651A (T1-2); 736 (S6-2); 738 (T1-2); 842 (S6-2); 957A (H7-1); 1150A (S6-2); 1485K (S1-1); 1574D (S7-4) – (U)

Wullschlägel, H.R., 1243 (S7-2a) type of *Trichosteleum martii* – (BR); s.n. (H3-2); s.n. (S5-6) – (NY)

FRENCH GUIANA

Andersson, L., 1350 (S6-2); 1354 (S5-6); 1360 (S6-2); 1363 (S6-2); 1367 (S6-2); 1368 (S6-2); 1372 (S5-6); 1373 (S6-2) – (NY)

Aptroot, A, 15190 (S5-5); 15216A (T1-2); 15252 (H7-1); 15299 (H7-1); 15300 (T1-3); 15325 (S5-5); 15403 (H6-1); 15445 (T1-2); 15446 (S5-6); 15454 (L1-2); 15456 (L1-2); 15513 (S2-1); 15543 (S6-2); 15556 (L1-2); 15567 (S6-3); 15578 (T1-2); 15580 (H2-1); 15582 (L1-2); 15584 (H7-1) – (U)

Bekker, J.M., 2287A (S1-1); 2309B (H7-1) – (U)

Benoist, R., s.n. (T1-1); s.n. (S6-2); s.n. (S5-6) – (NY)

Boom, B. & S. Mori, 1503p.p. (L1-2); 1510 (S5-6); 1523 (S6-2); 1526 (S7-4); 1542 (S7-4) – (NY)

Broadway, W.E., 392 (S6-2); 856 (S6-2); 866 (S6-2); 867 (S6-2) – (NY)

Buck, W.R., 18293 (S2-1); 18295 (S5-5); 18316 (S5-6); 18326 (S5-6); 18334 (S5-6); 18349A (S6-3); 18355 (S1-1); 18382 (S7-4); 18403 (S1-1); 18409 (S7-4); 18508 (15-1); 18514 (S7-4); 18577 (S7-4); 18585 (S1-1); 18628 (S7-4); 18637 (S5-5); 18674 (S5-5); 18695 (S2-2); 18704 (S5-6); 18735

(S6-3); 18750 (S7-4); 18771 (S5-5); 18812 (S7-4); 18911 (S7-4) – (NY)

Cornelissen, H. & H. ter Steege, 219 (S5-5); 259 (S5-5); 310 (S6-2); 319A (S5-6); 327 (H2-1); 343 (S6-2) – (U)

Cremers, G., 3814 (S6-2); 3837 (S7-4); 3840 (S7-2a); 3861 (S4-1); 3888 (S6-3); 3890 (S7-4); 3893 (S1-1); 3896 (S6-2); 4008 (S1-1); 4014 (H7-1); 4017 (H7-1); 4021 (H7-1); 4022 (H7-1); 4023 (S6-2); 4026 (H7-1); 4038 (S5-6); 4057 (L1-2); 4063 (T1-2); 4073 (H7-1); 4081 (T1-2); 4151 (H2-1); 4154 (H7-1); 4156 (S1-1); 4169 (H2-1); 4173 (S5-6); 4179 (H7-1); 4181 (T1-3); 4182 (H2-1); 4183 (H2-1); 4184 (H2-1); 4187 (H2-1); 4193 (S6-2); 4196 (H2-1); 4198 (H7-1); 4199 (H2-1); 4204 (H2-1); 4210 (S7-4); 4212 (H7-1); 4213 (H1-1); 4221 (H2-1); 4223 (S7-4); 4231 (S1-1); 4248 (S6-3); 4249 (S6-3); 4257 (S7-4); 4269 (L1-2); 4317 (S7-4); 4320 (S5-6); 4701 (S6-1); 4702 (S6-1); 4705 (H7-1); 4719 (S6-1); 4720 (S1-1); 4723 (S6-1); 4726 (T1-2); 4733 (H6-1); 4743 (S5-1); 4745 (S5-1); 4766 (L1-2); 4696 (S5-5); 4746 (S5-5); 4755 (S6-1); 4760 (S5-5); 4762 (S6-1); 4767 (S7-2a); 4771 (L1-2); 4781 (S6-2); 4784 (L1-2); 4785 (L1-2); 4786 (S5-5); 4791 (L1-2); 4800 (S6-1); 4808 (H7-1); 4877 (S5-6); 4884 (S5-6); 4895 (S6-2); 5131 (H6-1); 5131A (S7-3); 5133 (H2-1); 5135 (T1-2); 5137 (H2-1); 5149 (L1-2); 5153 (L1-2); 5144 (S6-1); 5156 (H2-1); 5160 (S6-1); 5162 (S6-1); 5167 (T1-2); 5168 (S5-5); 5170 (S6-1); 5173 (S3-2); 5186 (H7-1); 5224 (S5-6); 5290 (S5-6); 5298 (L1-2); 5303 (S6-3); 5316 (H6-1); 5317 (L1-2); 5323 (S5-6); 5349 (S7-4); 5449 (H6-1); 5451 (S7-4); 5456 (H1-1); 5458 (S6-2); 5469 (S5-6); 5481 (S5-6); 5561 (S7-4); 5617 (S6-3); 5630 (S5-6); 5644 (S6-3); 5673 (T1-2); 5674 (T1-1); 5796 (H2-1); 5798 (H2-1); 5809 (S1-1); 5811 (H2-1); 5817 (T1-2); 5818 (S5-6); 5820 (S7-4); 5821 (H2-1); 5833 (S6-2); 5861 (H1-1); 5858 (S5-6); 5865 (H1-1); 5867 (H1-1); 5869 (H2-1); 5872 (H7-1); 5882 (T1-2); 5887 (S3-2); 5893 (S5-6); 5894A (S6-3); 5896 (H6-1); 5905 (S6-2); 5906 (S6-3); 5912 (H6-1); 5921 (S5-6); 5975 (S1-1); 5976 (S7-4); 6025 (S5-6); 6200 (L1-2); 6204 (S5-6); 6209 (L1-2); 6210 (L1-2); 6228 (S1-1); 6229 (T2-2); 6232 (H6-1); 6257 (S5-6); 6259 (S7-3); 6291 (H3-1); 6304 (H1-1); 6307 (S7-2b); 6763 (S5-6); 6780 (L1-2); 6811 (S5-6); 6814 (L1-2); 6816 (S7-4); 6839 (T1-2); 6844 (H6-1); 6845 (H6-1); 6846 (S1-1); 6849 (L1-2); 6854 (L1-2); 6864 (S6-3); 6875 (H7-1); 6937 (S1-1); 7128 (S1-1); 7136 (S5-6); 7142 (S6-2); 7597 (S6-1); 7604 (S7-4); 7611 (H6-1); 7614 (S6-2); 7625 (S1-1); 7628 (L1-2); 7633 (L1-2); 7700 (H7-1); 8021 (H7-1);

8038 (T2-2); 8045 (L1-2); 8048 (T2-2); 8209 (L1-2); 8347 (H6-1); 8349 (S5-6); 8638 (S6-2); 8744 (S6-1); 8746 (T1-2); 8795 (L1-2); 8797 (H7-1); 8799 (H7-1); 8805 (T2-2); 8815 (H7-1); 8841 (S7-4); 8862 (L1-2); 8901 (H7-1); 8902 (H7-1); 8907 (L1-2); 8932 (T2-2); 8936 (H2-1); 8954 (S6-2); 8955 (S5-6); 8959 (H7-1); 9021 (S7-4); 9024 (L1-2); 9039 (H4-1); 9040 (H4-1); 9060 (S5-6); 9063 (H4-1); 9064 (H2-1); 9066A (H2-1); 9068 (H2-1); 9070 (S7-4); 9074 (H4-1); 9105bis (H2-1); 9106 (H6-1); 9116 (S1-1); 9130 (H7-1); 9136 (H4-1); 9166 (L1-2); 9167 (L1-2); 9190 (S5-6); 9215 (H4-1); 9220 (S7-4); 9224 (S5-6); 9243 (S8-1); 9251 (S5-5); 9301 (L1-2); 9308 (S5-6); 9320 (S3-2); 9990 (S5-6); 9991 (L1-2); 9997 (S5-5); 9998 (S3-2); 10023-10025 (S3-2); 10093 (S5-6); 10972 (S1-1); 10975 (S7-4); 11030 (S7-4); 11035 (S6-3); 11036 (S5-6); 11037 (T2-2); 11051 (S1-1); 11063 (S5-6); 11640 (S1-1); 11648 (S5-6); 12109 (H2-1); 12354 (S5-6); 12361 (S1-1); 12374 (S1-1); 12376 (S5-5) – (CAY, U)

Feuillet, C., 3896 (T2-2) – (CAY)

Florschütz-de Waard, J., 5854 (T1-3); 5856 (H2-1); 5861 (S6-2); 5869 (S1-1); 5871 (S5-5); 5873 (H7-1); 5876 (L1-2); 5885 (S7-4); 5886 (S5-6); 5887 (H2-1); 5894 (H6-1); 5895 (S1-1); 5896 (S7-4); 5897 (H6-1); 5910 (16); 5914 (L1-2); 5916 (S6-2); 5920 (H7-1); 5921 (L1-2); 5926 (L1-2); 5927 (H7-1); 5930 (S7-4); 5934 (H7-1); 5940 (T1-2); 5943 (H6-1); 5948 (S5-6); 5950 (H6-1); 5952 (T1-2); 5953 (H3-1); 5957 (S6-2); 5958 (H7-1); 5959 (T1-2); 5961 (S6-1); 5962 (S7-2a) – (U)

Gradstein, S.R., 5750 (S6-2); 5768 (S6-2); 5810 (L1-2); 5819 (H2-1); 5836 (S5-5); 5861 (L1-2); 5883 (L1-2); 5896 (H6-1); 6267 (S5-6) – (U)

Granville, J.J. de et al., 814 (T2-2); 9427 (H6-1); 9567 (H7-1) – (CAY, U)

Hoff, M., 5250 (H7-1); 5257 (H2-1); 5292 (S5-6); 5299 (S5-6); 5317 (H6-1); 5326 (H2-1); 5330 (S5-6) – (CAY, U)

Leprieur, M., hb. Montagne 7 (S3-2) type of *Potamium vulpinum*; 283 (S4-1); 304 (S1-1); 313 (S4-1); 1378 (S5-3) type of *Sematophyllum lonchophyllum*; s.n. (T1-2); s.n. (S4-1); s.n. (S5-6); s.n. (H2-1) – (PC)

Michel, A., s.n. (H6-1) ; s.n. (H2-1) type of *Ectropothecium guianae* – (L); s.n. (S7-3) – (NY)

Montfoort, D. & R.C. Ek, 471-473 (H1-1); 474-482 (L1-2); 483 (H7-1); 485 (H7-1); 486-492 (H6-1); 494-498 (H7-1); 499-500 (L1-2); 501 (H7-1); 502-507 (S2-1); 508 (H7-1); 509 (16); 514 (H3-2); 520-523 (S5-6); 524-525 (S7-4); 526-530 (S5-5); 531-536 (S1-1); 795A (S6-3); 795-808 (S6-2); 811-818 (T1-1) – (U)

Mori, S.A., 18511 (S6-2); 18513-15 (S6-2); 18659 (S6-2) – (NY)

Richard, C., s.n. (H2-1) type of *Ectropothecium leptochaeton* – (PC); hb Hooker 3181 (S5-6) type of *Hypnum richardii* – (BM)

Rey, gouv. M. (coll. M. Galliot), s.n. (H6-1); s.n. (H3-2) type of *Ectropothecium isopterygioides*; s.n. (S7-3) type of *Trichosteleum intricatum* ; s.n. (S2-1) – (PC)

Sagot, P., s.n. (S7-4) – (NY); s.n. (S3-2) – (H)

Sipman, H.J.M., 31623 (S6-2); 31832 (S6-3); 31833 (S1-1); 31842 (S6-2); 31849 (S7-4); 31853 (S5-6) – (B, U)

Villiers, J.F., 3396 (S5-6) – (PC)

INDEX TO SYNONYMS

Acanthocladium
    costaricense Dixon & Bartr. = S8-1
Acosta
    cuspidata C. Müll. = L1-2
Acroporium
    guianense (Mitt.) Broth. = S1-1
    intricatum Thér. = S7-3
Aptychus
    apaloblastus C Müll. = S5-5
    grammicarpus C. Müll. = S5-5
    leucodontaceus C. Müll. = S3-1
    micropyxis C. Müll. = S7-2a
Dimorphella
    pechuellii (C. Müll.) Ren. & Card. = H6-1
    trinitensis (C. Müll.) Herz. = H6-1
Donnellia
    commutata (C. Müll.) W.R. Buck = S2-2
Ectropothecium
    apiculatum Mitt. = H2-1
    globitheca (C. Müll.) Mitt. = H2-1
    guianae Broth. & Par. = H2-1
    isopterygioides Card. & Thér. = H3-2
    trinitense C. Müll. = H6-1
Fabronia
    donnellii Aust. = S2-2
Glossadelphus
    truncatulus (C. Müll.) Fleisch. = H5-1
Hookeria
    retusa Wils., nom. nud. = H5-1
    strumosum Hornsch. = L1-2
Hypnella
    jamesii Robins. = H5-1
Hypnum
    concavum Hook. = S6-1
    cuspidatifolium C. Müll. = L1-2
    diminutivum Hampe = H1-1
    galipense C. Müll. = S5-2
    globitheca C. Müll. = H2-1
    hornschuchii Hampe = S7-2a
    leptochaeton Schwaegr. = H2-1
    lonchophyllum Mont. = S5-3
    martianum Lor. = S7-2a

martii C Müll. = S7-2a
microcarpum Hornsch. = S7-2a
microtheca C. Müll. = S4-1
papillosum Hornsch. = S7-4
planum Brid. = S6-2
pungens Hedw. = S1-1
reptans Hedw. = H4-1
richardii Schwaegr. = S5-6
schistocalyx C. Müll. = T1-3
spirale C. Müll. = S7-4
strumosum (Hornsch.) C. Müll. = L1-2
subbrevisetum Hampe = H3-1
subsimplex Hedw. = S5-6
surinamense Dozy & Molk. = H7-1
tenerum Sw. = H3-2
trinitense C. Müll. = H6-1
truncatulum C. Müll. = H5-1
vesiculare Schwaegr. = H7-1
Isopterygium
radicisetum (C. Müll.) Broth. = H3-2
Leskea
caespitosa Sw. = S5-5
involvens Hedw. = T1-1
kegeliana C. Müll. = S5-5
subpinnata Brid. = S5-5
Leucomium
cuspidatum (C. Müll.) Jaeg. = L1-2
guianense C. Müll. = L1-2
Ligulina
octodiceroides C. Müll. = S5-3
Maguireella
vulpina W.R. Buck = S3-1
Meiotheciopsis
commutata (C. Müll.) W.R. Buck = S2-2
Meiothecium
negrense Spruce ex Mitt. = S3-1
tenerum Mitt. = S2-2
Microthamnium
diminutivum (Hampe) Jaeg. = H1-1
Mittenothamium
diminutivum (Hampe) Britt. = H1-1
Neckera
boryana C. Müll. = S2-1
commutata C. Müll. = S2-2

vulpina Mont. = S3-2

Plagiothecium
  radicisetum C. Müll. = H3-2

Potamium
  casiquiariense Spruce ex Mitt. = S4-1
  leucodontaceum (C. Müll.) Broth. = S3-1
  lonchophyllum (Mont.) Mitt. = S5-3
  octodiceroides (C. Müll.) Broth. = S5-3
  pacimoniense Mitt. = S5-4
  recurvifolium Thér. = S5-1
  uleanum Broth. = S5-3

Pterogonidium
  microtheca (C. Müll.) Broth. = S4-1

Pterogonium
  pulchellum Hook. = S4-1

Rhaphidostegium
  cochleatum Broth. = S5-1
  grammicarpum (C. Müll.) Par. = S5-5
  subdemissum Schimp. ex Besch. = S7-2a

Rhaphidostichum
  guianense Bartr. = S5-2

Sematophyllum
  apaloblastum (C. Müll.) W.R. Buck = S5-5
  caespitosum Mitt. = S5-5
  fluviale Mitt. = S7-2a
  guianense Mitt. = S1-1
  kegelianum (C. Müll.) Mitt. = S5-5
  maguireorum W.R. Buck = S3-2

Sigmatella
  guianae C. Müll. = S7-4
  quelchii C. Müll. = S6-1

Taxithelium
  patulifolium Thér. = S6-3
  quelchii (C. Müll.) Par. = S6-1

Thuidium
  acuminatum Mitt. = T. urceolatum Lor., name in note under T2-1, 2
  antillarum Besch. = T2-2
  delicatulum (Hedw.) B.S.G. var. *peruvianum* (Mitt.) Crum = T2-1
  involvens (Hedw.) Mitt. = T1-1
  scabrosulum Mitt. = T1-2
  schistocalyx (C. Müll.) Mitt. = T1-3
  verrucipes C. Müll. = T1-2

Trichosteleum
  fluviale (Mitt.) Jaeg. = S7-2a

guianae (C. Müll.) Broth. = S7-4
martii (C. Müll.) Kindb. = S7-2a
micropyxidium (C. Müll.) Broth. = S7-2a
pluripunctatum Ren. & Card. = S6-3
subdemissum (Besch.) Jaeg. = S7-2a
Vesicularia
surinamense (Dozy & Molk.) Broth. = H7-1

KEY TO THE GENERA OF MOSSES OF THE GUIANAS

bu

J. Florschütz-de Waard

The key is based on the characters of the taxa occurring in the Guianas and does not take account of characters of the species outside this area. Page numbers refer to the genera treated in the three parts of the moss flora. New additions to the flora will be treated in Musci IV; most of them have been included in the recent checklist of the Guianas (Florschütz-de Waard 1990, Boggan et al. 1992; see Introduction).

1 Branches arranged in fascicles on the stem. Leaf cells forming a regular network of wide hyaline cells enclosed by narrow chlorophyllose cells·· *Sphagnum* (pg. 22)
Branches not arranged in fascicles. Leaf cells not forming a regular network of two cell types· 2

2 Leaves with duplicate, sheathing bases, arranged in two opposite rows· *Fissidens* (pg. 28)
Leaves different · 3

3 Leaves with broad, sheathing bases and upper laminas with numerous longitudinal lamellae on the costa· · · · · · · · · · · · *Polytrichum* (Musci IV)
Leaves different · 4

4 Minute plants, less than 1 mm high, growing scattered from a persistent, algaelike substrate (protonema) · · · *Micromitrium* (*Nanomitrium* pg. 175)
Plants larger, protonema not persistent · 5

5 Plants whitish green. Leaf for the greater part composed of the costa, lamina reduced to the basal part of the leaf; costa in cross section with two or more layers of large hyaline cells (leucocysts) and one layer of small chlorophyllose cells (chlorocysts) · 6
Plants seldom whitish green (except *Bryum argenteum*). Leaf lamina well developed, costa various, if wide not occupying the whole leaf width· · · 9

6 Chlorocysts in cross section triangular; leaf apex flat, lingulate · *Octoblepharum* (pg. 106)
Chlorocysts in cross section four-angled; leaf apex tubulose or cucullate · 7

7 Costa with a longitudinal bundle of stereids from base to apex· *Leucophanes* (pg. 98)
Costa without a bundle of stereids · 8

8 Capsule inclined, curved. Leaf apex tubulose · · · · · · · *Leucobryum* (pg. 99)
    Capsule erect, cylindrical. Leaf apex more or less cucullate · · · · · · · · · · ·
    · · · · · · · · · · · · · · · · *Holomitriopsis* (*Leucobryum laevifolium* pg. 101)

9 Leaves without costa or with very short, usually indistinct, double costa · ·
    · Group C
    Leaves with a distinct costa, extending at least 1/4 of leaf length · · · · · · 10

10 Costa single · Group A
    Costa double · Group B

GROUP A. Genera with unicostate leaves.

1 Costa strong, reaching leaf apex · 2
    Costa well developed but vanishing above midleaf, not reaching leaf apex ·
    · 32

2 Leaf base with a conspicuous central group of wide, hyaline cells (cancellinae),
    usually clearly distinct from the smaller green upper lamina cells · · · · · ·
    · *Calymperaceae* (pg. 117)
    Basal cells of leaf base, if wide and hyaline, not forming a conspicuous
    central group · 3

3 Alar cells at leaf base differentiated, forming a distinct group · · · · · · · · 4
    Alar cells not differentiated, at least not in a distinct group · · · · · · · · · · 10

4 Alar cells coloured and/or inflated, often forming auricles · · · · · · · · · · · 5
    Alar cells small, quadrate, not coloured · 8

5 Costa wide, occupying 1/3 of leaf base or more · · · · · *Campylopus* (pg. 70)
    Costa narrower · 6

6 Leaves with a hyaline border of narrow, elongated cells, at least in the basal
    part · *Leucoloma* (pg. 83)
    Leaves without hyaline border · 7

7 Leaf cells linear and strongly incrassate throughout the leaf. Leaf apex with
    a hyaline hairpoint · · · · · · · · · · · · · · · · · · · *Eucamptodontopsis* (pg. 90)
    Only basal cells elongated and sometimes incrassate, upper leaf cells more
    or less quadrate. Leaf apex acute · · · · · · · · · · · · · · · *Holomitrium* (pg. 91)

8 Leaves narrowly lanceolate, folded lengthwise. Capsules immersed · · · · · ·
    · *Archidium* (Musci IV)
    Leaves ovate or oblong, concave. Capsules exserted. · · · · · · · · · · · · · · · 9

9 Alar cells rounded-quadrate, strongly incrassate, forming a large, conspicuous group. Plants often bearing microphyllous branchlets · · · · · · · · · · · · · · ·
· *Pseudocryphea* (pg. 243)
Alar cells quadrate or rhombic, little incrassate, not sharply separated from the surrounding cells. Leaves often arranged in spiral rows · · · · · · · · · · ·
· *Pireella* p.p.(pg. 249)

10 Plants with erect stems, simple or sparingly branched · · · · · · · · · · · · · · · 11
Plants with creeping primary stems (stolons) and with erect or prostrate secondary stems or branches. · 24

11 Upper leaf cells lax, elongate-hexagonal, over 10 μm wide · · · · · · · · · · · 12
Upper leaf cells quadrate, rectangular or linear, less than 10 μm wide · · · 16

12 Leaves with a border of narrow, elongated cells · · · · · · · · · · · · · · · · · · 13
Leaves with a border of quadrate cells or without distinct border · · · · · · 14

13 Capsule erect, endostome shorter than exostome · · · · · · · · · · · · · · · · · · ·
· *Brachymenium* (pg. 192)
Capsule inclined, endostome as long as exostome · · · *Bryum* p.p. (pg. 183)

14 Marginal leaf cells shorter, more or less quadrate in several rows. Apex obtuse, costa ending just before apex · · · · · · · · · *Splachnobryum* (pg. 179)
Marginal leaf cells not differentiated. Apex acute or rounded-acute and apiculate, costa percurrent or excurrent · 15

15 Leaves oblong or obovate-spathulate, 3-4 mm long · · · · · *Funaria* (pg. 177)
Leaves lanceolate, not over 3 mm long · · · · · · · · · · · *Bryum* p.p. (pg. 183)

16 Upper leaf cells isodiametric, quadrate or rounded hexagonal · · · · · · · · · 17
Upper leaf cells more elongate, rectangular to linear · · · · · · · · · · · · · · · 21

17 Leaf margin thickened coarsely double-serrate ·
· *Pyrrhobryum* (*Rhizogonium* pg. 200)
Leaf margin not thickened · 18

18 Leaf cells smooth (occasionally slightly papillose in *Hyophila*) · · · · · · · 19
Leaf cells densely papillose · 20

19 Leaves 1-2 mm long, apex acute-acuminate; upper leaf cells pellucid, over 10 μm in largest diameter · *Barbula* p.p. (pg. 169)
Leaves 1.5-3 mm long, apex obtuse to broad-acute, mucronate; upper leaf cells obscure, sometimes minutely papillose, less than 10 μm in largest diameter · *Hyophila* (pg. 171)

20 Leaves 2 mm or more long. Inner basal leaf cells wider and 4-6 times as long as upper leaf cells · *Trichostomum* (Musci IV)
Leaves not over 1.5 mm long. Inner basal leaf cells hardly wider, 2-3 times as long as upper leaf cells · *Barbula* p.p. (pg. 169)

21 Leaf cells smooth, irregularly rectangular · · · · · · · · · · · *Dicranella* (pg. 94)
   Leaf cells with papillose-projecting cell ends · 22

22 Plants slender, leaves less than 2 mm long · · · · · · · · · · *Philonotis* (pg. 202)
   Plants robust, leaves over 2 mm long · 23

23 Leaves linear-subulate with long-excurrent costa; cells in basal part of the
   leaf linear, smooth, in upper part oblong, strongly papillose by projecting
   apical cell ends · *Leiomela* (Musci IV)
   Leaves lanceolate with gradually acuminate apex; all cells linear, papillose
   at both cell ends · *Breutelia* (pg. 206)

24 Plants with erect, simple or branched secondary stems or branches · · · · · 25
   Plants prostrate · 29

25 Plants with distant, simple or branched, secondary stems often frondose or
   dendroid · *Pireella* p.p. (pg. 249)
   Plants with closely spaced, mostly simple branches, forming dense mats
   · 26

26 Inner basal leaf cells quadrate or transversely elongate, at margin bordered
   with elongated cells · *Groutiella* (pg. 207)
   All basal leaf cells elongate · 27

27 Leaves crispate or spirally twisted when dry. Calyptra conic, fringed at base,
   often plicate · *Macromitrium* p.p. (pg. 221)
   Leaves appressed when dry, only at end of branch sometimes twisted.
   Calyptra cucullate or campanulate · 28

28 Cells in basal leaf half linear. Calyptra cucullate (conic and split on one side)
   · *Macromitrium pellucidum* (pg. 232)
   Cells in basal leaf half rectangular or rhomboidal, often flexuose. Calyptra
   campanulate and lobed at base · · · · · · · · · · · · · · · *Schlotheimia* (pg. 215)

29 Leaves uniform, strongly asymmetric with a curved costa · · · · · · · · · · · 30
   Leaves di- or trimorphous, subsymmetric, dorsal leaves much smaller than
   lateral leaves · 31

30 Leaf cells rounded-quadrate, multi-papillose · · · · · · · *Mniomalia* (Musci IV)
   Leaf cells elongate-rhomboidal, smooth ·
   · *Phyllodrepanium* (*Drepanophyllum* pg. 197)

31 Leaves dimorphous, contorted when dry, lateral leaves oval, dorsal leaves
   triangular; costa long-excurrent · · · · · · · · · · · · · · · · *Racopilum* (pg. 236)
   Leaves trimorphous, strongly inrolled when dry, lateral leaves lingulate,
   dorsal leaves oblong, ventral leaves small, embedded in tomentum; costa
   percurrent · *Helicophyllum* (Musci IV)

32 Upper part of the leaf hyaline. Plants silvery white· · · · · · · · · · · · · · · · · · ·
· *Bryum argenteum* (*B. candicans* pg. 183)
Upper part of the leaf green · 33

33 Leaf cells papillose· 34
Leaf cells smooth · 39

34 Leaf cells with a small papilla at apical end· · · · *Porotrichum* p.p. (pg. 283)
Leaf cells with one or more papillae over the lumen· · · · · · · · · · · · · · 35

35 Leaf cells isodiametric. Plants pinnately or bipinnately branched · · · · · · 36
Leaf cells elongate. Plants irregularly branched· · · · · · · · · · · · · · · · · · 37

36 Plants slender, paraphyllia simple, leaf cells with small papillae on both
surfaces · *Cyrtohypnum* (pg. 372)
Plants medium-sized, paraphyllia branched, leaf cells with one or several
larger papillae on abaxial surface · · · · · · · · · · · · · · · *Thuidium* (pg. 379)

37 Leaf cells with 1 or 2 papillae on both surfaces. Alar cells incrassate,
quadrate, forming a conspicuous group ·
· *Henicodium* (*Leucodontopsis* pg. 242)
Leaf cells seriate-papillose on both surfaces. Alar cells short but not forming
a conspicuous group · 38

38 Leaf cells with 2-4 papillae. Leaves appressed when dry · · · · · · · · · · · · ·
· *Papillaria* (pg. 254)
Leaf cells with 3-8 papillae. Leaves erect-spreading when dry · · · · · · · · · ·
· *Floribundaria* (Musci IV)

39 Alar cells strongly differentiated in a conspicuous group · · · · · · · · · · · · 40
Alar cells, if differentiated, not forming a conspicuous group· · · · · · · · · 43

40 Plants with creeping stems and prostrate branches; leaves complanate or
erect-spreading, leaf apex acute or obtuse· 41
Plants with long, often pendent secondary stems; leaves regularly imbricate,
leaf apex abruptly acuminate or mucronate· 42

41 Leaf cells short, rhomboidal · · · · · · · · · · · · · · · *Stereophyllum* (Musci IV)
Leaf cells elongate· · · · · · · · · · · · · *Entodontopsis* (*Stereophyllum* pg. 352)

42 Branch leaves inserted in spiral rows. Alar cells in a flat, triangular or
quadrate group · *Orthostichopsis* (pg. 247)
Branch leaves not inserted in spiral rows. Alar cells in a round, inflated group
· *Squamidium* (pg. 257)

43 Branch leaves complanate. Cells in leaf apex rhombic or elongate-
rhomboidal · 44
Branch leaves imbricate or erect-spreading. Cells in leaf apex linear· · · · 45

44 Leaves strongly asymmetric or falcate, often transversely undulate. Apex obtuse or truncate ························· *Neckeropsis* (pg. 273)
   Leaves subsymmetric. Apex short-acuminate or mucronate ············
   ················ *Porotrichum* p.p. (pg. 283 and *Pinnatella* pg. 281)

45 Leaves bordered with several rows of elongated cells ··············· 46
   Leaves not bordered ······································· 47

46 Leaves oblong-lanceolate, acuminate ············· *Daltonia* (Musci IV)
   Leaves oval-oblong, apiculate ················· *Leskeodon* (Musci IV)

47 Creeping stems (stolons) with small scalelike leaves, secondary stems erect with broad-ovate, acute leaves, costa often variable, short and double or single ······································· *Jaegerina* (Musci IV)
   Secondary stems prostrate or pendulous; stem leaves ovate with long-acuminate apex ·· 48

48 Leaves 5 mm or more long, with a piliform acumen, as long as the leaf lamina ···························· *Spiridentopsis* (pg. 246)
   Leaves not over 2 mm long ································· 49

49 Secondary stem leaves with a strongly clasping base and a flexuose, piliform acumen ·············· *Zelometeorium* (*Meteoriopsis patula* pg. 259)
   Secondary stem leaves spreading from the insertion, acuminate ·········
   ·················· *Meteoridium* (*Meteoriopsis remotifolia* pg. 262)

GROUP B: Genera with bicostate leaves
(costae extending 1/4 of leaf length or more)

1 Leaf cells papillose··· 2
  Leaf cells smooth ··· 5

2 Leaf cells unipapillose ······································· 3
  Leaf cells multipapillose with 3-6 papillae in a row ··· *Hypnella* (pg. 295)

3 Plants with erect, frondlike secondary stems, pinnate- to bipinnate-branched ··· *Callicosta* (pg. 298)
  Plants with prostrate branches, complanate-foliate ··················· 4

4 Leaves lanceolate with a long flexuose apex ·· *Hookeriopsis* p.p. (pg. 311)
  Leaves ovate-oblong with rounded apex ····························
  ······················· *Callicostella* p.p. (*Schizomitrium* pg. 319)

5 Plants with erect or ascending branches or secondary stems ··········· 6
  Plants prostrate, leaves usually complanate························· 9

6 Plants with erect, frondlike secondary stems, pinnate- to bipinnate-branched
· *Callicosta* (pg. 298)
Plants with erect or ascending branches, sparingly divided · · · · · · · · · · · 7

7 Leaves linear-lanceolate, longitudinally plicate · · · · · · *Hemiragis* (pg. 304)
Leaves ovate, oblong or lanceolate, not longitudinally plicate · · · · · · · · · 8

8 Leaves transversely undulate in upper part · · · · *Hookeriopsis* p.p. (pg. 311)
Leaves not transversely undulate, contorted when dry· · · · · · · · · · · · · · · ·
· *Lepidopilum* p.p. (pg. 333)

9 Leaf cells large and lax, over 15 μm wide, along the margin a distinct
border of narrow cells · 10
Leaf cells narrower, leaves without distinct border · · · · · · · · · · · · · · · · 11

10 Costae slender, less than 40 μm wide at base. Margin subentire · · · · · · · · ·
· *Cyclodictyon* (pg. 306)
Costae firm, at least 40 μm wide at base. Margin dentate in upper part · · · ·
· *Lepidopilum* p.p. (pg. 333)

11 Leaves polymorphic, varying from ovate with obtuse, entire apices on stems,
to lanceolate with acute, dentate apices on distal parts · · · · · · · · · · · · · ·
· *Thamniopsis* (pg. 308)
Leaves uniform along stem and branches· 12

12 Costae extending about 3/4 of leaf length · 13
Costae extending less than 3/4 of leaf length (1/2-3/4 in *Lepidopilum
cubense*)· 14

13 Leaves more than 3 times as long as wide. Upper leaf cells oblong-linear
· *Hookeriopsis* p.p. (pg. 311)
Leaves less than 3 times as long as wide. Upper leaf cells isodiametric or
oval · · · · · · · · · · · · · · · · · · *Callicostella* p.p. (*Schizomitrium* pg. 317)

14 Leaves strongly concave with inflexed margins, apex blunt, often
canaliculate (leaf cells generally seriate-papillose but occasionally
nearly smooth) · *Hypnella* (pg. 295)
Leaves plane, apex acute or acuminate · 15

15 Seta smooth, capsule with transversely striate and furrowed exostome teeth.
Leaves with abruptly acuminate apex. Plants often reddish· · · · · · · · · · · ·
· *Lepidopilidium* (pg. 330)
Seta papillose, capsule with papillose exostome teeth without median furrow.
Leaves with acute or gradually acuminate apex · · · · · · · · · · · · · · · · · · ·
· *Lepidopilum* p.p. (pg. 333)

GROUP C: Genera with ecostate leaves.
(or with short and indistinct, double costa)

1 Leaf cells papillose or appearing papillose · 2
   Leaf cells smooth · 9

2 Leaf cells covered with a reticulum of many fine pits, appearing papillose;
   marginal cells smooth, forming a distinct border · · *Rhacocarpus* (pg. 241)
   Leaf cells papillose; leaf not bordered · 3

3 Leaf cells papillose only by projecting cell ends · · · · · · · · · · · · · · · · · · · 4
   Leaf cells with papillae over the lumen · 5

4 Leaf cells projecting at distal ends only · · · · · · *Mittenothamnium* (pg. 451)
   Leaf cells projecting at both ends · · · · · · · · · · · · · *Chrysohypnum* (pg. 441)

5 Leaf cells with one papilla · 6
   Leaf cells with more papillae · 7

6 Leaf apex pungent, with involute margins. Inflated alar cells large (70-170
   μm long), curved to the insertion · · · · · · · · · · · · · · *Acroporium* (pg. 386)
   Leaf apex flat or with partly revolute margins, not pungent. Inflated alar cells
   not over 100 μm long · *Trichosteleum* (pg. 424)

7 Papillae spiny, irregularly scattered over the cell lumen and often at cell
   ends; leaf apex truncate with coarsely dentate margin · · · · · · · · · · · · · ·
   · *Phyllodon* (pg. 453)
   Papillae arranged in rows over the cell lumen; margin smooth or slightly
   serrulate · 8

8 Papillae multifid, stalked · *Hypnella* (pg. 295)
   Papillae low and round · *Taxithelium* (pg. 417)

9 Plants with elongated and often pendulous secondary stems · · · · · · · · · · 10
   Plants with creeping or floating stems, usually freely branched and forming
   dense mats · 14

10 Stem and branches strongly flattened, often attenuate. Leaves arranged in 2
    or 4 opposite rows · 11
    Stem and branches not flattened. Leaves strongly concave, with broadly
    inflexed margins near apex · 12

11 Leaves falcate to sickle-shaped in 4 rows; apex flat, acute · · · · · · · · · · · · ·
    · *Isodrepanium* (pg. 278)
    Leaves concave-cymbiform in 2 closely imbricating rows; apex boat-shaped,
    apiculate · *Phyllogonium* (pg. 264)

12 Branch leaves not conspicuously ranked, apex with long, flexuose acumen. Leaf base strongly auriculate, with a small, round group of coloured alar cells · *Renauldia* (Musci IV)
Branch leaves ranked in spiral rows, apex apiculate. Leaf base not auriculate, alar cells few in a poorly defined group · 13

13 Branch- and stem leaves spirally ranked; medium-sized plants, branch leaves 1-2 mm long · *Orthostichidium* (pg. 246)
Only branch leaves spirally ranked; slender plants, branch leaves to 1 mm long · *Pilotrichella* (pg. 253)

14 Plants growing periodically submerged; stems and branches often elongate, sometimes floating; leaves usually with rounded apex (acute in *Sematophyllum lonchophyllum*) · 15
Plants not typically growing submerged; stems creeping with branches erect or prostrate · 18

15 Leaf cells uniform, elongate-hexagonal, alar cells not differentiated. Capsule immersed · 16
Leaf cells at midleaf at least twice as long as in apex, alar cells differentiated, the basal row usually inflated. Capsule exserted · · · · · · · · · · · · · · · · 17

16 Leaves densely imbricate particularly towards ends of branches. Capsule with single peristome · *Hydropogon* (pg. 238)
Leaves distant, flaccid. Capsule without peristome *Hydropogonella* (pg. 239)

17 Leaves oval or semi-circular, to 1.4 mm long. Capsule with slender, fragile peristome; exostome teeth not transversely striolate on outer surface, endostome reduced to filiform segments, often rudimentary · · · · · · · · · · ·
· *Potamium* (pg. 394)
Most leaves longer than 1.4 mm. Capsule with firm peristome; exostome teeth transversely striolate on outer surface, endostome well-developed, with a high basal membrane and broad, keeled segments · · · · · · · · · · · ·
· *Sematophyllum* p.p. (pg. 402)

18 Alar cells conspicuously differentiated in a well-defined group, at least in lateral leaves · 19
Alar cells not or scarcely differentiated, not forming a conspicuous group
· 26

19 Alar cells quadrate or rectangular, in the basal row sometimes oval, but not inflated · 20
Alar cells in the basal row distinctly inflated · · · · · · · · · · · · · · · · · · · 22

20 Leaves asymmetric, oval-oblong with obtuse or broad-acute apex, alar cells only differentiated in one leaf edge of the lateral leaves · · · · · · · · · · · · ·
· *Pilosium* (pg. 358)
Leaves symmetric, ovate-lanceolate with acute apex, alar cells differentiated in all leaves in both leaf edges · 21

21 Leaf cells thin walled, margin serrulate········· *Pterogonidium* (pg. 399)
   Leaf cells incrassate with a fusiform lumen, margin entire ············
   ································· *Meiothecium* p.p. (pg. 389)

22 Leaves on complanate branches dimorphous: on dorsal side broad-ovate,
   usually falcate, on ventral side lanceolate or triangular, subsymmetric···
   ································· *Rhacopilopsis* (pg. 456)
   Leaves uniform, branches complanate or not ····················· 23

23 Leaf apex with involute margins, pungent. Inflated alar cells large, 70-170
   µm long··································· *Acroporium* (pg. 386)
   Leaf apex plane or partly reflexed. Inflated alar cells not over 100 µm long
   ··· 24

24 Plants pinnately or bipinnately divided, branches attenuate, often curved.
   Branch leaves smaller than stem leaves ············· *Wijkia* (pg. 435)
   Plants irregularly branched, branches erect or prostrate. Branch leaves
   usually equal to stem leaves, sometimes larger··················· 25

25 Leaf margins narrowly reflexed from just below the lingulate apex to the
   base. Peristome single, exostome teeth slender, pale, widely spaced ····
   ································· *Meiothecium* p.p. (pg. 389)
   Leaf margins flat or loosely reflexed. Peristome double, exostome teeth
   thick, brown, endostome well-developed with a high basal membrane and
   broad, keeled segments················ *Sematophyllum* p.p. (pg. 402)

26 Branches / secondary stems erect or ascending, leaves squarrose or erect-
   spreading ··· 27
   Branches prostrate, leaves complanate ························· 28

27 Leaves erect-spreading, narrowly lanceolate with slender, serrulate apex ··
   ································· *Lepyrodontopsis* (Musci IV)
   Leaves squarrose-spreading, broad-ovate with acute apex (costa often
   variable, short and double or single) ··········· *Jaegerina* (Musci IV)

28 Midleaf cells lax, elongate-rhomboidal, over 10 µm wide············· 29
   Midleaf cells linear, less than 10 µm wide ····················· 30

29 Midleaf cells to 70 µm long. Peristome teeth not furrowed ···········
   ····································· *Vesicularia* (pg. 459)
   Midleaf cells 70-240 µm long. Peristome teeth with a median furrow ····
   ····································· *Leucomium* (pg. 366)

30 Branch leaves apparently in 4 rows····························· 31
   Leaves in more than 4 rows ······························· 32

31 Leaves on complanate branches dimorphous: on dorsal side broad-ovate, usually falcate, on ventral side lanceolate or triangular, subsymmetric · *Rhacopilopsis* (pg. 456)
Leaves variable, only inserted at dorsal side of the stem, at ventral stem side dense clusters of rhizoids. Epiphyllous plants· · · *Crossomitrium* (pg. 291)

32 Plants pinnately branched. Leaves usually falcate and strongly homomallous. Perichaetia conspicuous, inner perichaetial leaves erect, to 4.5 mm long ·
· *Ectropothecium* (pg. 444)
Plants irregularly branched. Leaves subsymmetric to slightly falcate, more or less complanate-spreading. Perichaetia not conspicuous, inner perichaetial leaves patent, to 1.5 mm long · · · · · · · · · · · · · · · *Isopterygium* (pg. 446)